Cover image by Adam P. Hayes

Questions and comments are encouraged and you can send them to
TrueBeautyOfMath@gmail.com

THE TRUE BEAUTY OF MATH

Volume 1

The Foundations

Contents

	Introduction To The Series	1
	How To Read The Series	6
	Note On Notation	9
1	**Sets**	**12**
	One True Sentence	12
	Elements And Sets	15
	Exercises	19
2	**Subsets**	**20**
	The Definition	20
	The Empty Set	23
	Something Concrete	24
	Exercises	31
3	**Sets Of Sets**	**32**
	The Basic Idea	32
	Power Sets	34
	Exercises	37
4	**A Paradox**	**38**
	Here Comes Trouble	38
	A Warm-Up	41
	The Main Event	43
5	**Functions**	**48**
	Getting Some Intuition	48
	Notation And Terminology	53

	Exercises	55
6	**Injective, Surjective, Bijective**	**56**
	Adding Structure	56
	Injectivity	57
	Surjectivity And Bijectivity	61
	Exercises	65
7	**Infinity And Hotels**	**66**
	Dealing With Infinite Sets	66
	The Hilbert Hotel	69
8	**Bijectivity And Counting**	**77**
	A Return To Injectivity And Surjectivity	77
	The Most Inefficient Way To Count	80
9	**Counting Infinity**	**83**
	A Problem With Our Counting	83
	Fixing The Problem	85
10	**Infinity Times 2**	**88**
	Being Precise About Infinity	88
	Infinity Times 2	89
	Infinity Type 1 Plus Anything	92
	Exercises	94
11	**Infinity Times Infinity**	**95**
	Setting The Stage	95
	A Quick Refresher	96
	A Seemingly New Infinity	98
12	**Infinity Times Infinity = Infinity**	**102**
	Needing To Get Creative	102
	Finding The Function	104
13	**Proof By Contradiction**	**111**
	Axioms	111
	Statements And Contradictions	113

Example: Infinitely Many Primes 117

14 Infinity Type 2   125
In Between Fractions 125
The Working Man's Definition Of Real Numbers 129
Cantor's Diagonal Slash 132

15 Infinity Infinities   139
Contrapositives 139
Which Infinity Is Bigger? 146
The Next Infinity 148
Infinity Infinities 153
Parting Thoughts On Infinity 156

16 Unions And Intersections   160
How We Approach Mathematical Structures 160
Two Important Constructions 162
Exercises 168

17 Playing Well With Others   169
A Whole New Class Of Questions 169
Unions Of Unions 171
Intersections Of Intersections 176
Putting It All Together — 3 Sets 179
Exercises 182

18 More Constructions, Part 1   184
The Cartesian Product 184
A Look Towards The Future 187
Exercises 189

19 More Constructions, Part 2   190
The Disjoint Union 190
An Important Subtlety 193
A Pause In Construction 196
Exercises 197

20 Reconsidering High School Algebra	**198**
The Foundations	198
A Fresh Perspective On Planes	201
A Fresh Perspective On Lines	207
21 The Necessity Of Irrationality	**214**
A Bit More Notation	214
A Chain Of Proper Subsets	216
The Proof	223
Outro And Intro	**230**
Solutions To Exercises	**232**
Solutions To Chapter 1	232
Solutions To Chapter 2	237
Solutions To Chapter 3	240
Solutions To Chapter 5	246
Solutions To Chapter 6	255
Solutions To Chapter 10	262
Solutions To Chapter 16	263
Solutions To Chapter 17	267
Solutions To Chapter 18	275
Solutions To Chapter 19	280
Appendix A: Vacuous Truths	**298**
Appendix B: Proving The Pattern	**301**
Index	**307**
Index Of Notation	**308**

Introduction To The Series

Mathematics is the crown jewel of the human intellect. It is the language that we use to describe and understand the natural world around us on every scale and it is the tool that we use to model almost all human activity. Most impressively, mathematics is the subject that allows mankind to transcend the physical world by uncovering beautiful and eternal truths about an abstract realm constructed through pure logic.

Yet mathematics remains one of the most misunderstood and polarizing subjects of all intellectual endeavors. Those who love math do so for very deep, almost spiritual reasons, while almost everyone else quite literally has nightmares about the subject. Very few people today have any idea at all about the scope of mathematics, the essential ideas and modes of reasoning that underly mathematics, or what is truly involved in actually "doing" mathematics. The truth is that there is an entire world of mathematics out there that most people live and die without ever getting to see.

The blame for this widespread misunderstanding of — and hatred towards — mathematics is hard to allocate. It is not the fault of math teachers, for math has just as many bad teachers as any of the subjects that are not afflicted by this widespread hatred. More positively, the subject of mathematics has fighting in its corner an army of smart, passionate, and dedicated math teachers, as well as an extensive academic literature, all trying to understand and change the status quo in math education. Nor does the blame fall on math students, for students that despise math usually have a right to do so, largely due to the fact that the math that they learn is inherently dull.

In fact I do not believe that the blame for the current popular hatred of math rests on **who** is teaching or learning it, but rather on **what** it is that is being taught and learned. Almost all of the math that is taught at the middle school and high school levels is several centuries old. Since most people do very little math after high school, there are few who have any idea how diverse and expansive the mathematical world truly is. Almost everyone is aware that **music**, for example, can exist in seemingly infinitely many different guises. It is common knowledge that there is classical music, jazz, rock 'n' roll, hip-hop, country, and several other immensely diverse genres with all of their hundreds of sub-genres. Yet the only mathematics that most people know exists are the very specific (and rather dull) subjects of high school algebra, high school geometry, and maybe some calculus. Mathematical fields like category theory, differential and symplectic geometry, algebraic topology, group theory, representation theory, combinatorics, number theory, and all of their hundreds of sub-fields, form an immense and beautiful chunk of human intellectual endeavor, yet most people have no idea that they even exist.

Let us take this musical analogy even further in order to clarify the current state of mathematical understanding. Imagine a society in which musical scales (major scales, minor scales, or any other kind of musical scale) were the only type of music that one could ever play or listen to. No one could ever hear music from artists like Bach, Beethoven, Justin Timberlake, Michael Jackson, Miles Davis, Garth Brooks, or Jay-Z. Coffee shops, bars, and night clubs could only put on scales, and any aspiring musician had to spend at least 10 years practicing nothing but scales on their instrument of choice.

I would venture to guess that in such a backward society there would develop a general hatred of music, since most people would be wrongfully equating the entire craft of music with the act of playing scales. In other words, most people would think that music **was** scales, and therefore conclude that they hate music as a whole. And this widespread hatred would be warranted — there is very little enjoyment that can come from playing or listening to scales.

There would, of course, be a small subset of the population who would find a unique kind of purity and simplicity in musical scales, and only this small portion of the populace would happily wade through the required 10 years of scale practicing. Suppose that once this 10-year landmark is reached, this small portion of music students got to go on to study the music of Wagner and Ellington and The Rolling Stones and Bruno Mars. Their love of music would deepen dramatically as they will have started to expose themselves to a wonderfully diverse world of music, and they would start to see how all of their scale practicing fits in to a much larger and more beautiful world of melody, harmony, and rhythm.

This lucky few who happened to make it through the initial 10-year scale requirement would undoubtedly be viewed as crazy radicals by their non-musical peers, since the latter would not have any ability to understand why it is that anyone would want to devote their life to such a difficult and unsatisfying craft. In this nonsensical world, professional musicians spend their time studying and playing beautiful and intricate music, yet no one else in this society has the ability to interact with this music because in their schools and coffee shops and nightclubs the only thing that plays are musical scales. Non-professionals would have the idea that professional musicians spend their days practicing and playing boring old scales, when in reality professional musicians are interacting with an art form that is unimaginably diverse and gorgeous.

I claim that the above and seemingly absurd scenario is precisely the situation that we find ourselves in today with regards to math and math education. The math that we learn for about the first ten years of our education is solely what I call "mathematical scales." These are the aspects of math that can be broken down into a set of instructions and memorized in order to solve problems of a very particular and limited variety. Much of this type of math is extremely important to learn (as are musical scales), but it is hundreds of years old and relatively dull. More importantly, this type of math is an awful representative of the types of reasoning and thought processes that are truly at the core of mathematical exploration. Only those who find some joy in the austerity and objectivity of these "mathematical scales" stick with the craft long enough to see this other world of mathematical beauty and diversity, yet their interests are completely misunderstood by the general populace who believes that mathematicians spend their days "solving for x."

While this is all of course a minor annoyance for the mathematician herself,[1] it is more importantly an impossible state of affairs for a craft if that craft ever desires to obtain popularity. And if it ever becomes important to **society** to have this craft's popularity rise,

---

[1] We will not assume that we know the gender of our reflexive or third person pronouns, and therefore we will bounce back and forth between using "herself," "himself," "her," and "him," so as to also avoid the more clunky verbiage of "her/himself," "her/him," and "s/he."

then this state of affairs must be turned around. Indeed, we now find ourselves in a society in which the modes of reasoning inherent to mathematics are becoming more and more sought after, while much of the population has no idea that these modes of reasoning even exist.

It is therefore necessary to begin to change this status quo, and to make the true nature of mathematics and mathematical exploration more widely understood. In order to do this, let us consider how we do this in the musical setting. In music it is indeed **not** the case that we require our musicians to first spend 10 years studying nothing but scales before they get to play some actual songs. Instead, we allow our aspiring musicians to play some easy songs **while** practicing some scales. Moreover, we are rather lenient with our students' scale practicing in their early years. This is because in these early years it is most important that the **love** of music is developed in the student, and this is developed by playing songs, not scales. Once this love for the craft is developed, **the student will motivate himself to practice scales, as he understands that it benefits him to do so, because it improves his capabilities at a craft that he enjoys**.

I am therefore not claiming that we need to stop teaching the (very important) fields of high school algebra, high school geometry, and calculus. What I am claiming is that there is no amount of academic literature, nor any sufficiently fun iPad or computer game, that can help people see the true nature of mathematics when these are the only subjects we study. These subjects are in large part inherently dull, with their main redeeming qualities being their austerity and objectivity. As long as these subjects are the sole contributors to the math curriculum of our young math students, it will continue to be the case that only those students who appreciate this austerity make it through to see the "other side" of math, and it will continue to be the case that there is a widespread misunderstanding of what math can offer the world.

In this series of books my goal is to introduce the reader to the equivalent of the "easy songs" of mathematics. In other words, my goal in *The True Beauty Of Math* is to provide a resource for going beyond the standard middle school and high school curriculum, and

INTRODUCTION TO THE SERIES

for going beyond the mathematical scales. In this way I aim to make a small contribution towards rectifying the current widespread misunderstanding of and hatred towards math. There is no mathematical background that the reader is assumed to have — these books are designed to be rather self-contained. All that is asked for of the reader is a willingness to accept the possibility that there is a whole new world of math out there. This new world of math is very different from the world that we have seen in middle school and high school, and we will be creating this world from the ground up, using nothing but logic. If the reader can take these facts on faith for now, then we are ready to dive in and start seeing for ourselves the true beauty of math.

I will warn you, though, that you might end up really loving math.

How To Read The Series

The new world of math that we will begin exploring in this volume is very different from the world that we have seen in middle school and high school, and therefore it is unavoidable that some concepts will appear somewhat difficult upon a first reading. We will be creating this world from the ground up, using nothing but logic, and in such a way that each new construction builds upon constructions that came before it. Therefore, it is likely that some paragraphs (or even chapters) will have to be read more than once, but more importantly, it is very likely that the reader will have to do a very great deal of thinking. In particular, this means that the reader will often find herself staring up from the page and blankly out into space, biting her nails and/or tapping her feet, deep in thought. This is good. This is necessary.

Mathematics requires thought, and lots of it. Deep thought can be (and often is) difficult, since it shakes us out of our routine thought processes, which are easy. However, mathematics is **beautiful**, and it is our desire to **experience** this beauty that motivates us to carry out this deep and sometimes difficult thought. Throughout these volumes I do my best to lessen the amount of thought required by giving as clear and as thorough of explanations as I possibly can, but this will never be completely successful. Therefore it is most important when reading this series to give oneself the time, and find within oneself the energy, to devote whatever amounts of thought are required to fully understand each and every step of each and every chapter. One good way to do this is to not only mull over each paragraph in the main body of the chapters, but to also spend time with the exercises at the end of (some of) the chapters, to give them a thorough attempt yourself, and to deeply understand the solutions to those exercises (which are provided in the back of each volume).

These volumes are not "page turners" in the same sense that many popular novels are. These volumes are meant to expand and stretch the types of thought processes that we are used to by exploring mathematical structures and concepts that were likely foreign to the reader before opening this book. Reading these volumes therefore requires more active participation on behalf of the reader than reading most

HOW TO READ THE SERIES 7

books does, but there is also more to gain.

I strongly recommend first reading the "Introduction To The Series" above (if this has not already been done) and then reading the "Note On Notation" below. The latter will help minimize one's very natural fear of mathematical symbols, of large equations, and therefore of math itself. I also strongly recommend repeatedly returning to the Note On Notation any time one feels overwhelmed by the symbols and notation used in mathematics, since at the end of the day (and as the Note On Notation describes) doing mathematics is no different than reading (and seeing as you are understanding the letters on this page I am assuming that reading is a skill that can be taken for granted here). After the Note On Notation, each chapter is meant to be read sequentially, and the solutions to any chapter that has exercises are meant to be an integral part of the logical content of this work. Certain results, ideas, and definitions that are necessary for progressing through each volume (including the current volume) are discussed in the solutions, and therefore the solutions should be viewed as an equally important aspect of these volumes as the main text itself. The volumes themselves can be read either sequentially, or in any way such that the prerequisite volumes (which will be listed at the beginning of any given volume) are read before the given volume is begun. The first volume — the current volume — is a prerequisite to all volumes.

Before diving in to this volume, I want to make it clear that none of the mathematical results discussed in this series are original. Namely, I am not laying out a plan for how I **personally** believe that math **should be**, nor am I making a **proposal** for some kind of "new mathematics." Instead, what I am doing in this series is sharing and describing results that are **already well-established** amongst the mathematical community, most of which have been established since long before I or any reader of this series were born. Thus, I am not presenting my own personal ideas of how math **should be**, but rather I am simply communicating how math **already is**. Many of the examples and analogies are indeed original (as far as I know), but the **mathematical content itself** is well-established and truly fundamental. This note is necessary only because the math that we will be learning in these volumes will likely be wildly unfamiliar. It is therefore important to establish that this new world of math upon which we are about to gaze is **not** the product or opinion of a single person but rather the result of centuries of hard work by well-trained mathematicians — we just have yet to see it. Until now.

Note On Notation

One reason why many people are scared of or even hate mathematics is that once people see an expression like

$$: A_0(x_1) \cdots A_0(x_n) := \sum_{X \subset \{1,\ldots,n\}} \prod_{i \in X} A_+(x_i) \prod_{j \in CX} A_-(x_j)$$

they run and hide. Let us step back for a moment and try to see what is really going on here, so that we can better understand this fear, and possibly do away with some of it.

The above expression is scary not because the ideas inherently hidden within it are bad or ugly, but rather only because we do not understand the notation of the expression. In other words, we do not understand the chosen symbols used to **represent** the ideas. It might as well be Egyptian hieroglyphics for all we know. In fact, math **could** have been written down in Egyptian hieroglyphics and its logical content would still be exactly the same. Why is this? To answer this question, we need to better understand the role that notation plays for us in mathematics. In doing so, we will understand one of the many reasons why math is in fact not so scary.

Let us begin with a question: What do we mean when we write down the number 3?

Note that we are **not** asking "what is the number 3?", but rather we are asking what the meaning of the symbol "3" is. This is a subtle but extremely important issue. The symbol 3 is shorthand notation for the abstract concept of "three" which we keep in our heads. Thus, when I write down 3 and when you read 3, I can know that we are keeping the same abstract idea in our minds. In other words, the symbol 3 means nothing without the abstract meaning that we give it. Moreover, it is simply a shorter, more concise way of keeping track of the abstract notion of "three-ness." We could have written any other symbol in its place, and its abstract meaning would be the same so long as that is what we agreed on.

Let us take an example that is even closer to home. Consider, for example, the words on this page. You are looking at this strange concoction of squiggly and straight lines, interspersed with blank white space, and you are deriving from it certain meaning. Simply put, you are reading. You are extracting meaning from the abstract ideas that are represented by a relatively arbitrary mixture of symbols. Moreover, the meaning derived from these symbols is independent of the symbols themselves. In other words, we could have all agreed to use other symbols and the abstract ideas would be left intact, so long as we agreed on what the symbols meant. Therein lies the creation of new and different languages!

In addition to representing words (which are abstract ideas) by symbols on a page, we can define new words based on old words. In doing so, we can have shorter ways of expressing more and more meaning. For example, we defined the word "bus" so that we could encapsulate all of the meaning of "bigger version of a car, designed for transporting large amounts of people" without having to always write that whole long phrase down. Thus, when we read the word "bus" we automatically associate it with this larger meaning, and the word "bus" is just a convenient abbreviation. We then can use this word to define a new word (or in this case, a new phrase), namely "school bus." This way we can simply say "school bus" instead of always having to say "bigger version of a car, designed for transporting large amounts of people to and from school."

Math is no different. We use symbols to wrap up more and more meaning into less and less writing. This has no effect on the logical content of a given mathematical statement or equation — notation is simply used so that we humans can communicate mathematics to each other more efficiently. Thus, if we can read then we are already doing exactly what we might have been scared to do by looking at the above equation — namely, we are deriving abstract information from a language designed to express and summarize certain ideas.

Here is a quick way to become less scared of math. Consider the term "symplectic manifold." There is a decent chance the reader has no idea what a symplectic manifold is, and therefore is confused, and therefore is worried that he will never understand math, and therefore is scared of math forever, and therefore tries to learn guitar instead, and therefore becomes that guy at parties always playing the guitar and pissing people off. I claim that there is a better way.

As mentioned before, the words "symplectic manifold" are just abbreviations for other constructions, which are in turn abbreviations for other constructions — just like "school bus" (and just like the symbols in the expression at the beginning of this section — each standing for some larger, well-constructed idea). In fact, a symplectic manifold is nothing but a differentiable manifold with some additional properties, and a differentiable manifold is nothing but a manifold with some additional properties, and a manifold is nothing but a topological space with some additional properties, and a topological space is nothing but a set with some additional properties, and we study sets in the first chapter of this volume, requiring no further background! Of course I am not trying to say that a symplectic manifold is a particularly easy thing to study, but I **am** saying that there is nothing to be inherently scared of. If we can make it through Chapter 1 of this volume, then we are well on our way to learning what a symplectic manifold is. In both a literal and a metaphorical sense, then, all of mathematics is now at our fingertips...

Chapter 1

Sets

One True Sentence

In this chapter we will introduce two of the most fundamental ideas in all of mathematics. Virtually all of mathematics can be built up from logical deductions made from what we describe in the next few pages. But before we introduce these ideas it is important for us to take a moment to ask ourselves what we even mean by "mathematics," and the qualities that we want this subject to have.

There have been entire libraries written trying to fully understand the question "what is mathematics?", so we will not attempt to contribute to that literature here. Instead, let me motivate our discussion by using (as I tend to do) a fictional universe as our guide. Suppose we lived in a universe where you were one of the first intelligent beings of your species, so that no notion of mathematics existed yet. Suppose also that you loved to argue with your fellow species members and suppose you hated being wrong. Throughout the course of one of your days you find yourself in an argument with one of your friends about whether or not $2 + 2 = 1$. You insist that indeed sometimes $2 + 2 = 1$, while your friend insists that this can never be the case. You lie awake for several nights, fuming with frustration about your lack of ability to convince your friend that you are right.

After several nights of frustration you realize that there is only one way that you can convince your friend that you are right, and to do so in such a way that he could **never** argue with or doubt your result. What you aim to do is to state one fact — just one single statement — that your friend agrees with. Once that is done, all you need to do is take that fact and logically deduce another fact from it (if your friend cannot accept logical deductions, then I would recommend getting new friends). Now that you have arrived at a new fact that your friend agrees with, you can then logically deduce another fact from this second fact. You have now arrived at a third fact, which was logically deduced from the second fact which was logically deduced from the initial fact that your friend agrees with. If you continue on in this way, logically deducing each fact from the previous one, and if it is eventually the case that the fact that you land on is the statement that $2 + 2 = 1$, then you will be able to tell your friend **with certainty** that your statement is correct.

As you lie awake in bed thinking about how you will finally be able to prove your friend wrong, you see that this all boils down to being able to successfully do two things. You must first find a fact (or several) that your friend will be **absolutely forced** to agree with. Once you find this set of facts, you must then make a set of logical deductions such that your friend **absolutely agrees** with each step. And as you lie awake thinking of these things, you start to get rather ambitious. You now ask yourself whether or not there could be a set of facts that **everyone** would be **absolutely forced** to agree with — not **just** your friend. You then also ask whether or not it could be the case that there is a mode of reasoning that **everyone** would be **absolutely forced** to acknowledge the validity of.

In this alternate universe, your insatiable desire to prove your friend wrong has allowed you to create what we now call mathematics. Namely, mathematics is the craft that unambiguously derives new incontrovertible truths from previously established incontrovertible truths, using a mode of reasoning that is itself incontrovertible. It is believed that there is a set of incontrovertible truths that lie at the foundation of this whole story, but as we will see in a few chapters this is actually a rather subtle issue. Regardless, we will now begin creating mathematics by asking ourselves what the most incontrovertible truths are, and deducing new truths **from** these truths using nothing but pure logic. In doing so, we will actually see that in a particular sense the alternate universe version of yourself is right — there are situations in which $2 + 2 = 1$, but to prove this[1] we need to construct a setting in which discussing such topics is possible, and we call this setting mathematics...

So what is the most fundamental, incontrovertible truth that we can think of? One thing that might come to mind is the following. To me, "the sky is blue" is a pretty fundamental truth, one that most people can agree on. However, we quickly find that this statement relies on the fact that the sky exists, that we are talking about the sky on planet Earth, and also that we all know what blue is. This is therefore not all that fundamental of a truth. Let us instead try "Kobe is better than LeBron" as our fundamental truth.[2] For a Laker fan like myself, this seems like a pretty good fundamental truth to start building all of math upon, but unfortunately it has some faults. The statement is indeed true but it is not at all fundamental. Namely, the statement "Kobe is better than LeBron" rests on the fact that we all know what "better" means (which can be tricky), and more importantly it rests on the fact that both Kobe and LeBron even exist. In fact, we see that every truth that we could ever write down will run into the problem of first needing to assume that something exists, and therefore the truth that we wrote down is not fundamental (since it relies on some other fact).

Luckily this problem points us in the direction of its own solution. Namely, we must take as our fundamental truth the statement that "something exists." That is it. That is the only statement that we must all believe to be true, and that is the statement on which we will build almost all of mathematics. But let us pause for a moment to appreciate the fact that this is indeed as fundamental of a truth as we can possibly think of right now. The only alternative to this truth would be the statement that "nothing exists," and a theory of mathematics built up in a universe in which nothing exists would be a boring theory indeed.

---

[1] We will indeed prove this explicitly in the next volume of this series, so let us read on to get there!

[2] For those amongst us who are not NBA fans, Kobe and LeBron two of the most famous basketball players of our generation and are regularly used as examples throughout this volume.

Elements And Sets

We now are all in agreement that "something exists," and this is actually a tremendous amount of progress. We have not specified precisely **what** exists, nor precisely in which condition these things may exist. All we have agreed to is that indeed **something** exists. Let us now give a name to things that exist, and let us call them **elements**. Namely, an element is simply a thing. Any thing. An apple is an element, a number is an element, an idea is an element, this book is an element. You are an element, I am an element, "humanity" is an element, and "democracy" is an element. Elements are **the most fundamental objects in all of mathematics**.[3]

Once we know that an element exists, we can suppose that there may be **several** elements in existence, and this supposition is the basis for the second most fundamental idea in all of mathematics. Namely, a collection of elements is a **set**. A set is just some elements. Thus, if we have an apple, a water bottle, an idea, and the number 3, we can take them all together and form the set whose elements are "an apple, a water bottle, an idea, and the number 3." We can consider the set of all living people in the world, or the set of all dead American presidents, or the set of all deodorant sticks. To summarize, an element is a thing, and a set is a collection of elements. Thus, a set is just a collection of things.

We might now be wondering why we are talking about sets and elements — what do deodorant sticks and apples have to do with math? Our curiosity would be warranted, but it is unjustified. Remember, we are building mathematics **from the ground up** right now and we therefore cannot ask whether or not something "has to do with mathematics." We are creating math as we go! Right now all we have said is that mathematics is the study of sets and elements, and therefore it just so happens to be the case that apples and deodorant sticks and dead presidents all have to do with math. Of course,

---

[3]The experts reading this will know that this is not strictly true, for rather subtle reasons that we will completely disregard in these early volumes. If we were to start exploring the intricacies of "set theory proper" at this early of a stage then it would go fully against the aims of these volumes, as laid out in the introduction.

we can consider sets of **numbers** and get a little closer to what we are used to thinking math is all about, but we must remember that apples and presidents form just as good of sets as do numbers.

In order to make progress, we need to be a little more precise and we need to establish some notation. In the following, recall from the above "Note On Notation" that notation is nothing to be scared of. The way that we will build up our mathematical universe from our fundamental truth is by first making precise definitions of objects or ideas that we know exist, and then proving things about the objects that we define using nothing but agreed upon logical deductions. So let us get into the math spirit by making our first mathematical definitions (which are really just summaries of what we have discussed so far in this chapter).

Definition 1.1. An **element** is a thing.

Definition 1.2. A **set** is a collection of distinct elements.

We are sweeping a whole lot of subtlety under the rug with these definitions, and we will see what kind of trouble this gets us into in Chapter 4. For now, though, we continue on with them because they are the simplest and most intuitive definitions of an element and a set.

As an example, the book on my shelf, the water bottle on my desk, and the sun, are all elements. We can therefore consider the set whose elements are "the book on my shelf, the water bottle on my desk, and the sun," since this is now a collection of elements. Similarly, the book in your hands, the water bottle on your desk (if there is one), and the sun, are all elements, and so we can therefore consider the set whose elements are "the book in your hands, the water bottle on your desk, and the sun." We can now ask whether or not these two sets are the same set. In order to do so, we must first acknowledge the fact that we do not even know yet what it means for two sets to be the same, so we can hardly ask this question until we specify what "two sets being equal" even means! Remember that this is mathematics, and we therefore can **only** use modes of reasoning that we are **forced**[4] to agree on. With some thought, it becomes

---

[4]Precisely what it is that forces us to agree on these matters is rather mysterious, even possibly spiritual depending on your philosophy.

ELEMENTS AND SETS   17

clear that it **must** be the case that two sets are the same (or "equal") if **each one of its elements** are the same. Thus, the two sets that we defined at the beginning of this paragraph are **not** the same. The element "the sun" is likely the same between the two sets, assuming that you are reading this book in the same solar system that I wrote it in. But the other two elements in the first set, namely "the book on my shelf" and "the water bottle on my desk", are **different** elements than the other two elements in the second set ("the book in your hands" and "the water bottle on your desk"), and therefore these two sets are **not** equal.

There are more subtleties that we need to clear up, but before we do this let us set up some notation. If we are dealing with a set, we can label the set however we like. Thus we can say, "let A be the set of all basketballs," which would mean that every time we write the letter A, we are just using a shorthand notation — an abbreviation — for the abstract idea of "the set of all basketballs." Accordingly, if I were holding a basketball in my hands right now, we could truthfully say that that basketball is an element of A, which is short for "it is an element of the set of all basketballs."

Let us now continue developing some notation. When labeling the elements of a set, we put the elements inside brackets such as $\{\}$. Thus, $A = \{\text{all basketballs}\}$ is shorthand notation for the phrase "A is the set of all basketballs." There is nothing fancy going on here, it is just notation. Suppose we decided to call the set of all basketballs $BALL$ instead of A. Then we would have the abbreviation $BALL = \{\text{all basketballs}\}$. We could have instead decided to call this set $DONKEY$ if we wanted to, but that would not be very helpful since the elements of the set are basketBALLs, and not donkeys.[5]

Another common set that we will see in the future is the set of all positive whole numbers, namely 1, 2, 3, 4, and so on, forever. We will often denote this set either by $\{\text{positive whole numbers}\}$ or by $\{1, 2, 3, ...\}$ where the "..." is to remind us that this set "goes on forever." If we wanted to call the set of positive whole numbers C, then we would write $C = \{\text{positive whole numbers}\}$ or $C = \{1, 2, 3, ...\}$, and they would both mean the same thing. Remember, what we write

---

[5] However, there is nothing **inherently wrong** about calling this set $DONKEY$ — it is just a **preference** not do so.

on the page is just a bookmark for an abstract idea in our head. Therefore as long as we remember what the bookmark stands for, we will be alright. By our above definition of C, we have that 17 is in C (i.e., 17 is an element of C), but negative 5, denoted by -5, is not in C, nor is the water bottle on my desk (because neither of them are positive whole numbers).

Using this notation, we can clarify two more subtle issues before moving on to Chapter 2. Let us consider the following two sets, calling one of them A and one of them B: $A = \{1, 5, 2, 6\}$ and $B = \{1, 2, 6, 5\}$. Are these two sets the same? I.e., does $A = B$? The answer is yes, since they both have the exact same elements. Namely, every element that is in A is also in B, and every element that is in B is also in A (notice that both of these statements have to be true in order for the statement "A and B have the exact same elements" to be true). The point is that the **order** in which we write the elements does not matter, because the set has no "structure." In other words, when we are writing these sets down with the above notation, we are just listing the elements that are in the sets. Obviously, whether or not an element is in a set is independent of the order in which we write them, and all that matters for determining the equality (or inequality) of two sets are the elements that make up those sets.

Lastly, let us consider the following two sets (again calling them A and B): $A = \{1, 1, 2, 2, 7\}$ and $B = \{1, 7, 2\}$. Are these sets the same? **Yes.** Why? Well, we already know that the order in which we write the elements does not matter, but now we also recall that in the definition of a set (i.e., Definition 1.2 above), the elements in a set need to be **distinct**. Since 1 is not distinct from 1, and 2 is not distinct from 2, the set $\{1, 1, 2, 2, 7\}$ is really the same as the set $\{1, 2, 7\}$, and therefore $A = B$!

This might all seem rather trivial or obvious for us now (or it might not), but we will soon be seeing that these issues are in fact highly non-trivial. In fact, in a couple of chapters we will derive a real, full-fledged, inescapable paradox within this sort of set theory! For now, this gives us a new way of looking at mundane objects around the house or office or classroom, by combining them in various ways to form sets.

Exercises[6]

1) Define a set, any set at all, in two different ways. For example,

$$\{\text{all humans over the age of 35}\}$$

is the same set as

$$\{\text{all humans who are not 35 years of age or younger}\},$$

yet these two sets are defined in different ways. Similarly,

$$\{\text{those who think that LeBron is better than Kobe}\}$$

is the same set as

$$\{\text{those who enjoy being wrong}\}.$$

2) Try to define some crazy sets, and try to determine if two sets are the same or not. For example, can we define the set of all ideas? What about the set of all thoughts? What makes two different thoughts distinct? (Remember, elements in a set need to be distinct!) Is the set of all ideas equal to the set of all thoughts? It depends on how we define them! Let us not get too worked up about this stuff yet though — we are just trying to see that "mathematical thought" is indeed much broader than we could have ever imagined!

3) Is there a set that contains itself as an element? (This one is tricky and it is something we will be addressing more in Chapter 4, so we should not panic if this exercise eludes us right now.)

---

[6] Note that full and detailed solutions to all of the exercises are in the back of this volume.

Chapter 2

Subsets

The Definition

In Chapter 1 we defined, and became comfortable with, the notion of sets. Our discussion of sets might have seemed relatively trivial and/or obvious, and in no way related to math. But alas, this **is** math. We have been **creating** math as we go, and we have found that math is the study of sets![1] Additionally, these ideas that may seem like trivialities will lead us into some deep trouble that we will see in Chapter 4, and this will show us that these ideas are in fact highly non-trivial. The solution to the third exercise of Chapter 1 is closely related to this trouble, and we will address it in due time. Thus, we must continue to pay attention, even to the trivialities!

We now turn our attention to a new mathematical idea. In particular, let us try to understand the notion of **subsets**. Subsets are exactly what they sound like — they are "sub" other sets. If we are given some set — call it A — then a subset is just another set whose elements are all in A. In other words, a set B is a **subset** of another set A if it is the case that every element of the set B is **also** in the set A. The idea here is simply that, for example, the set $\{a, b, c\}$ is a subset of **any** other set that also contains (at least) the elements a, b, and c. Note that this definition precisely aligns with our intuitive notion of the meaning of one thing being "sub" another thing. In

---
[1] Well, at least **a lot** of math is.

THE DEFINITION

particular, if each of the elements of one set is also in some other set, then the former set is in a very precise sense "contained in" the latter set, just as we would want from a "sub" object. Let us turn this into a formal definition.

Definition 2.1. Given a set called A (we can call it whatever we like), a subset of A, call it B, is a set such that every element in B is also an element in A.

The best way to get used to any definition is to take a look at some examples, so let us do this now with the definition that we just made. Suppose we had the set $\{1, 2, 3, 4, 5\}$, whose elements are the whole numbers from 1 to 5. In this case, the sets $\{1, 2, 3, 4\}$, $\{2, 4, 5\}$, and $\{5\}$, are all subsets of the initial set, because each of these three sets satisfy the requirement that "each of their elements is also an element of $\{1, 2, 3, 4, 5\}$." Note that there are lots of other subsets of $\{1, 2, 3, 4, 5\}$ that we have not written down. Also note that the set $\{1, 2, 4, 6\}$ is **not** a subset of $\{1, 2, 3, 4, 5\}$ because the former has the element 6 in it, whereas the latter does not. Thus, the set $\{1, 2, 4, 6\}$ does **not** satisfy the requirement that "each of its elements is also an element of $\{1, 2, 3, 4, 5\}$," and therefore it is not a subset of $\{1, 2, 3, 4, 5\}$. For more examples of finding subsets of specific sets, see the first exercise of this chapter below.

Let us now discuss a few important points about subsets in general. First of all, any subset B of any set A is itself a set. Namely, the subset B is a perfectly well defined collection of elements in its own right.[2]

Another extremely important point is that for any set A, A is a subset of itself. Why is this? Well, by the definition of a subset, a subset is just some set whose elements all lie in some other set. Moreover, all of the elements in A certainly lie in A, so A is a subset of itself! There are two things to note here. First, we note that this is

---

[2] Note that we are calling sets and subsets A and B, and that this is completely arbitrary. We are simply giving a label to "any set" and "any subset of that set" just so that we can more easily reference them. The set A could be absolutely any set in the world, and the subset B could be absolutely any subset of that set. Moreover, we could have chosen to use (for example) the **symbol** "$DONKEY$" instead of A, and/or "PIG" instead of B, but hopefully we can agree that our choices are indeed simpler.

a perfectly logical construction — there is no reason why A **cannot** be a subset of itself. The second noteworthy fact is that we were **forced** to arrive at this conclusion (namely, that any set is a subset of itself) once we made the definition of a subset. In other words, we simply **defined** what we meant by "subset" and then were **forced**, logically, to conclude that every set is a subset of itself. We could of course change our definition of subset and be led to other conclusions, but **this** definition is actually very useful. This illustrates a large part, if not all of what mathematics is. Namely, it is the formulation of precise definitions and the study of the inevitable logical conclusions to which they lead.

Let us now turn the above paragraph into a formal theorem so that we can start to get the hang of how mathematics works. Namely, we have made some definitions, and from these definitions we have been able to prove a new statement (that every set is a subset of itself). Mathematics then boils down to three stages — making a definition, stating a theorem (i.e., a truth) about those definitions, and finding a proof of that theorem using previously established theorems. Unfortunately (or fortunately, depending on your philosophy), none of these three stages are easy. It is usually difficult to determine which definitions will be the most interesting definitions to study, it is then usually difficult to determine what is true about these definitions, and it is usually difficult to figure out how to **prove** those truths from previously established truths. Indeed, no one ever said that math would be easy, but we are here to see that it is **worth it**.

We have already made a couple definitions, so let us now state our first theorem.

Theorem 2.2. *Every set is a subset of itself.*

We give this theorem a number and its only little space on the page just because we will end up deriving more truths from it in the future, so we want to be able to easily reference it. We will also not explicitly prove it now, because we already have done so a couple paragraphs ago.

The Empty Set

Let us now discuss an extremely important set, known as "the empty set." The empty set is just exactly what it sounds like — the set that is empty. Recall that the definition of a set was just a collection of distinct elements. Well, the empty set is a perfectly good collection of distinct elements — it is the collection of no elements!

For now, and for a while, the notion of an empty set might seem pretty absurd. In particular, we might wonder why it would ever be interesting to consider a set with absolutely no elements in it. This is a completely justified question at this point in time, however it turns out that the existence of the empty set actually leads to a tremendous amount of logical beauty and consistency. We will see some of this beauty of the empty set in these volumes (including this first volume), and it continues to be an important concept in even higher mathematics. Thus the empty set proves to be a notion that we cannot do away with. We can draw whatever philosophical conclusions we would like from this, but no matter what, the empty set is real!

Let us now state one of the most important facts about the empty set. Namely, a little thought will show us that **every set has the empty set as a subset**. We can see this by remembering what it takes for one set to be a subset of another. Suppose we are given a set A (any set, any set at all) and let us ask whether or not the empty set is a subset of it. Upon reflection of the definition of a subset, we see that we need it to be true that every element in the empty set is also an element of A. Well, **there are no elements in the empty set**, so we can truthfully say that "there is no element in the empty set that is not also an element in A," simply because there is no element in the empty set to begin with! Since there is no element that is both in the empty set and not in the set A, we must conclude that every element in the empty set is also an element in A. We have therefore concluded that the empty set is a subset of A. However, since A is **any set at all** (namely, we never had to specify what A is in any way), we have shown that the empty set is a subset of every set.[3]

---

[3]This is an example of a **vacuous statement**, and for more on vacuous statements (which are hugely important in mathematics), see Appendix A.

The above was (possibly) the most confusing paragraph of this book so far, so according to the "How To Read This Book" section, a second or third reading of this paragraph may be necessary before moving on, and that is entirely okay.

Something Concrete

Let us get our heads out of the clouds with all of these abstract[4] sets and subsets and instead do something concrete. In particular, let us try to answer the following particular question: given a set with a finite number of elements in it, how many distinct subsets does it have? In other words, if the enemy handed us a set (and for reasons that we will see below, we need this set to have a finite number of elements, as opposed to infinitely many elements), we want to know how many different subsets we can make from it.

Before answering this question let us first ask **why** we might even want to ask this question in the first place. In particular, suppose we wanted to convince someone that this question is worthwhile to devote one's attention to. We could begin by giving the **economic or physical application** of this question, and along these lines we could explain how answering this question would allow us to understand the number of possible ways to distribute a collection of 10 apples or a million dollars or what have you. One could imagine several different scenarios in which this information would be directly and physically applicable to the society in which we live.

However, I claim that there is a deeper and ultimately **more important** reason for exploring this question. The more important reason for asking this question is simply that it is **interesting**. As we will see shortly, the answer to this question is simple and beautiful.

---

[4]By "abstract" we simply mean some idea that does not require having a **particular** example in mind. In particular, much of our discussion has involved **any** set at all, and/or **any** subset of any particular set. The idea is that if we can prove a statement about **any** set at all (i.e., if we can prove a statement about the abstract idea of a set), then that statement will automatically be true for any **particular example** of a set. This is the power of mathematics. Of course, in order to get a better understanding of a particular abstract idea, it is always best to take a look at some specific, concrete examples of that abstract idea.

Additionally, the **logic** that we will use to uncover this result is beautiful and important in its sheer existence. The physical or economic utility of the solution to the problem is simply a lucky coincidence — a fortunate happenstance in which human society can benefit from mathematical endeavor. There is no doubt that such happenstances are useful and important to uncover, but it is also important to distinguish mathematical exploration from its practical application. As we discussed before, mathematics is the study of well-defined structures and the truths to which they inevitably lead using pure logic. It is purely a fortunate occurrence that such intellectual endeavors lead to practical applications, but this is **not** the actual **nature** of mathematics. Therefore as mathematicians, we move towards those questions that are **interesting** in their own right, and this question is one such question. Let us see why.

Let us first investigate why this question is posed the way that it is. In particular, why do we limit ourselves to finite sets, as opposed to allowing for the possibility that our set has infinitely many elements? The answer is that we limited ourselves in this way so that we can get a finite number of distinct subsets. In particular, if we ask how many distinct subsets an infinite set has, we will always get an infinite number. The reason we would get infinitely many distinct subsets if our "starting set" was infinite is that we can always just form a 1-element subset from each element in the set. Thus if there are infinitely many elements in our set, there will be infinitely many such 1-element subsets — and these are not even all of the subsets!

For example, suppose we want to know how many distinct subsets there are of the set

$$A = \{\text{positive whole numbers}\} = \{1, 2, 3, 4, 5, ...\}.$$

Then $\{1\}$ is a subset of A (since every element in the former is also an element in the latter), $\{2\}$ is a subset of A, $\{3\}$ is a subset of A, and so on. We clearly get infinitely many subsets by forming these 1-element subsets, and this is not even considering subsets like $\{1, 2\}$, or $\{5, 18, 29\}$, or $\{45, 67, 12763\}$. Thus, if we want to get an actual finite number (as opposed to infinity) as our answer to the question of "how many subsets of a particular set are there?", then we better limit ourselves to asking about sets with only finitely many elements.

Finally, why do we ask for **distinct** subsets of a given set? Suppose we consider the set $A = \{1, 2, 3\}$. Then the set $\text{SET}_1 = \{2, 3\}$ is a subset of A, as is $\text{SET}_2 = \{1, 3\}$, where we (arbitrarily) chose the names SET_1 and SET_2 for the sets $\{2, 3\}$ and $\{1, 3\}$, respectively. We also have that the set $\text{SET}_3 = \{2, 3\}$ is a subset of A. But $\text{SET}_1 = \text{SET}_3$! Thus SET_1 and SET_3 do not make distinct subsets, despite the fact that they both make subsets. We clearly want to know about the number of **distinct** subsets of a given (finite) set, since simply **renaming** the **same** subset over and over again (which is what we have done with SET_1 and SET_3 above) is not particularly interesting. And now that we fully understand **why** the question is asked in precisely the way that it is, let us finally tackle this problem.

Let us first get some familiarity with the problem by checking out some explicit examples.[5] Accordingly, let us first consider a set with only two distinct elements in it. Let us call the set A and its elements a and b, so that we have $A = \{a, b\}$. If we would like, we can suppose that a and b are anything — numbers, horses, apples, ideas, or whatever else. All that is important to us is that there are two distinct elements in this set.

We now ask the following question. How many distinct subsets of A are there? Well, we know that the empty set is a subset of A, since we proved before that the empty set is a subset of **all** sets. This makes one subset of A so far. We also know that A is a subset of A, since we also proved that any set is a subset of itself (this was Theorem 2.2). This makes two subsets of A so far. Next, we see that $\{a\}$ is a subset of A, and $\{b\}$ is a subset of A. This makes four distinct subsets of A. Moreover, we claim that this is **all** of the distinct subsets of A. To see this, we note that since A itself only has two elements, the subsets of A must have either 0, 1, or 2 elements.[6] There is only one 0-element set in the whole logical world,[7] and this is the empty set. There is also only one 2-element subset of A, because that **must** be A itself. And since A is a 2-element set, it can only have two 1-element subsets.

---

[5] Examining explicit examples is almost always the best way to begin to understand some general statement.

[6] A set cannot have a negative number of elements, and a subset of a set cannot have **more** elements than its "parent" set.

[7] See Exercise 3 of this chapter for a proof of this statement.

Thus counting up the number of 0-, 1-, and 2-element subsets of A, we see that there is a **total** of four distinct subsets of A. Since A is **any** 2-element set at all, we have therefore proven that there are 4 distinct subsets of any two-element set.

Let us now consider an arbitrary 3-element set, and let us similarly call it $A = \{a, b, c\}$. Let us count our distinct subsets by first counting the number of 0-element subsets (which we already know to be only one, namely, the empty set), then counting the number of 1-element subsets, then counting the number of 2-element subsets, and finally counting the number of 3-element subsets (which we also already know to be one, namely, the set A itself). We therefore have as our subsets the empty set (total current subset count = 1), the whole set A (total current subset count = 2), the three 1-elements subsets $\{a\}, \{b\}, \{c\}$ (total current subset count = 5), and the three 2-element sets $\{a, b\}, \{b, c\}$, and $\{a, c\}$ (total current subset count = 8). These are all of the possible subsets of $A = \{a, b, c\}$, and so the total number of distinct subsets of a 3-element set is 8.

We now also note that for a 1-element set, there are only two distinct subsets — the empty set and the set itself. Thus we have the following pattern, where the number of elements in the set are on the left and the number of distinct subsets of the set is on the right:

# of elements in the set	# of distinct subsets of that set
1	2
2	4
3	8

There is a pattern here — namely, an "exponential" one. Let us therefore remind ourselves what exponentials are all about. Recall that 2^3 is simply shorthand notation for "2 raised to the power of 3," or in other words "3 products of 2," so that $2^3 = 2 \times 2 \times 2 = 8$. Similarly, $2^4 = 2 \times 2 \times 2 \times 2 = 16$. Well, the pattern that we see forming above is that the number on the right is simply "2 raised to the number on the left." In other words, $2^1 = 2$, so in this case it is indeed true that "2 raised to the power of the number on the left of the first row (which is 1), is the number on the right of the first row (which is 2)." Additionally, $2^2 = 4$, so again we have that "2 raised to the power of the number on the left in the second row is equal to the number on the right of the second row," and similarly $2^3 = 8$, so that the same statement holds for the third row.

The pattern does indeed continue, and one can prove that a set with N elements has 2^N distinct subsets, so long as N is any whole number greater than or equal to 0. Note that this does indeed work for $N = 0$. We see this as follows. According to the pattern, a 0-element set should have $2^0 = 1$ distinct subsets (anything to the power of 0 is **defined** to be 1). And indeed, there is only one 0-element set, and that is the empty set. To see this we note that the empty set has the empty set as a subset, since **every** set has the empty set as a subset. But the empty set also has itself as a subset, since **every** set has itself as a subset. But "itself" **is** the empty set! Thus these two subsets are actually the same, and since we only count **distinct** subsets, we only need to include one of these sets! Moreover, the empty set cannot have any 1-, 2-, or 3-element subsets since the set itself has 0 elements. Therefore we have proven that the empty set has only one distinct subset (itself), just like the pattern predicted!

Note that we have not **proven** that the statement "any set with N elements has 2^N distinct subsets" is true for any non-negative whole number N — all we have done so far is prove that this statement is true for $N = 0, 1, 2,$ and 3. The general proof of this statement requires some mathematics that we will not see for a few more chapters. In particular, we will need to understand what a **function** is before we can give the general proof of this statement for any N, and we will not examine functions until Chapter 5. The ambitious reader is encouraged to try to either prove the general statement for any N on her own, or at least convince herself of its validity. There are a couple of different ways to prove it, and we can see one such way in Appendix B.[8]

Note also that this pattern (the pattern that a set with N elements has 2^N distinct subsets) relies heavily on the fact that the empty set is a subset of every set, and that every set is a subset of itself. These two facts about subsets came immediately from our definition of subsets. Namely, based on our definition of subset, we were **forced**

---

[8]Though this appendix should not be read until we have reached the end of Chapter 8, as well as understood Chapter 5 and its exercises (in particular, Exercise 4).

to conclude that every set is a subset of itself, and that every set has the empty set as a subset. Additionally, once we made our definition of subset, we were **forced** to conclude that a set with N elements has 2^N distinct subsets (though we have not proved this in full generality yet).

We could indeed **modify** our definition of subset in such a way as to not "pick up" the subsets "the set itself" and "the empty set" in our counting. For example, **suppose** we had decided to **define** a subset of a set A as follows: "if A is a set, then a subset of A is any set B that is **non-empty**, not A itself, and such that every element of B is an element of A." This is precisely the same definition of subset that we have in Definition 2.1 except for the added requirements that B is not empty, and is not A itself. We are therefore **manually** removing the empty set and the set itself from our counting of subsets, and in this case our pattern would be that a set with N elements has $(2^N - 2)$ subsets. This is simply because we would get all of the subsets that we got from our counting using our initial definition, except we will have "lost out" on precisely two subsets (the empty set and the set itself).

We also note that in this case (with this new definition of subset), the pattern of a set with N elements having $2^N - 2$ elements breaks down when $N = 0$, since this pattern would say that a set with N elements has -1 distinct subsets (since $2^0 - 2 = 1 - 2 = -1$). Since we can never have a negative number of subsets, this pattern simply does not work when $N = 0$, namely, for the empty set. Thus, not only do we get a slightly less elegant pattern with this definition of subset (simply because the pattern that "a set with N elements has 2^N subsets" is simpler than the pattern that "a set with N elements has $2^N - 2$ subsets"), we also see that the new pattern does not apply to as many values of N, since it does not work for $N = 0$, while our original pattern (from our original definition of subset) does indeed work for $N = 0$.

Similarly, we could just as well change our definition of subset in such a way as to only "lose out" on the empty set, by defining a subset of A to be a non-empty set B, all of whose elements are in A (note that this **does** allow A to be included as a subset of itself). Then we would find that a set with N elements has $2^N - 1$ subsets. There are clearly infinitely many different (and increasingly strange) definitions that we could **in principle** make, however, the definition of subset that we started with gives us a very simple and nice relationship between the number of elements in a set and the number of subsets of that set ($N \to 2^N$). Because this pattern is simple and elegant, and because many other even more beautiful results (that we will see much later in this series) come from the specific definition that we made, we will stick with our initial definition, Definition 2.1.

Let us emphasize that this pattern ($N \to 2^N$) is again just the **inevitable** logical consequence of precise definitions that we have made. We are at liberty to make whatever definitions that we want, but we are **not** at liberty to decide which consequences those definitions lead to — we are **forced** by the eternal and infallible rules of logic into whatever consequences our definitions yield. This is analogous to the following. We can take a piece of paper and draw a circle on it, and a line on it. We have total freedom as to where we draw the circle, how big the circle is, and where we draw the line. Once we have finished drawing, we can start asking questions about what we have drawn. Does the line intersect the circle? What angle is the line at relative to the bottom of the paper? If the line does not intersect the circle, how close does it get? While we were at liberty to draw whatever we wanted, we are now **not** at liberty to determine the answers to these questions. The answers to these questions are set in stone in some eternal world of absolute truth **once we have drawn our picture**. We could draw a **different** circle and line, and then we would be **stuck** with the ensuing results in exactly the same way. This is (a small part of) the beauty of mathematics.

Before concluding this chapter, let us reflect on the power of all this abstraction with sets and subsets. Namely, why is it so useful to consider abstract sets as opposed to always considering "concrete" sets like "a specific set of dogs" or "a specific set of NBA players"? The power of abstraction lies in the fact that if you have 10 dogs and I have 10 ideas and my friend has 10 cars, **all of us** now know how many ways we can "split up" these sets into subsets **regardless** of what these sets actually contain. This is because we have seen (though have not proven yet) that **any set at all** with 10 elements in it will have $2^{10} (= 1,024)$ distinct subsets. This is just one of many ways in which abstraction is powerful, and we will see more examples of this power and beauty in what is to come!

Exercises

1) (a) Write down four subsets of the set

$$\{a, b, c, d, e, f, g\}.$$

(b) Write down three subsets of the set

$$\{\text{Kobe}, \text{LeBron}, 8, \text{love}, \text{fruit loops}, 27\}.$$

2) If C is a subset of B, and B is a subset of A, then C is a subset of A. True or false?

3) It is hopefully clear that there are infinitely many 1-element sets "out there." This is because we can form the sets $\{1\}$, $\{2\}$, $\{3\}$, $\{4\}$, $\{5\}$, and so on, so this already gives us infinitely many 1-element sets (and this is not even including {cup}, {table}, {this chair}, {that chair}, { that chair over there}, {that other chair}, {lion}, and so on). Similarly, there are infinitely many 2-element sets, and infinitely many 3-element sets (and so on). The question is, then, how many 0-element sets are there? (Hint: We give away the answer somewhere in this chapter, so now we need to understand **why** this is so.)

Chapter 3

Sets Of Sets

The Basic Idea

In this chapter we study the notion of "sets of sets." This concept is an extremely powerful mathematical tool, but it is also the tip of a very deep and subtle mathematical iceberg. In particular, although the idea of "sets of sets" is remarkably useful in the development of mathematics, it can also be used to cause some serious problems regarding the foundations of mathematics (if we are not careful). In this chapter we will first explore the idea of a set of sets and then we will go on to start exploring how it can be problematic. In the next chapter we will see one of these problems first hand, and we will briefly describe how these problems are dealt with in practice. For now, we simply focus on the idea of sets of sets without any worry about possible future problems.

So what is a set of sets? As usual, it is exactly what it sounds like — a set of sets is a set whose elements are themselves sets. This may or may not seem a bit strange and/or too abstract, so let us take this slowly because it is extremely important. We begin by recalling what we need to make a set. Namely, all we need is a collection of distinct "things" which we call elements. As discussed before, elements could be numbers, or donkeys, or ideas, or some combination of any of these and anything else that we can think of. Accordingly, we can think of a set **itself** as a **single** element of some other set. After all, a set (as a whole) is a "thing," and we have a well-defined notion of when two sets are distinct from each other.[1] Therefore we are perfectly well able to consider sets whose individual, indivisible elements are entire sets. Let us look at some examples to see how this works.

Our first example will seem kind of tedious and unnecessary, but it will truly illustrate the power of the idea of "sets of sets," and it will hopefully make clearer the distinction between elements, sets, subsets, and sets of sets. Let us consider the following two sets: $SET1 = \{1, 2, 3, 4, 5, 6\}$ and $SET2 = \{\{1, 2\}, \{3, 4\}, \{5, 6\}\}$. Note the nested brackets in $SET2$. Here we are using the set notation that we developed in Chapter 1, and the brackets are nested in this way because we see that the **elements** of $SET2$ are themselves sets. In particular, if we had a set $A = \{a, b, c\}$ and we knew that the **element** a was the **set** $\{1, 2\}$, so that $a = \{1, 2\}$, then we could equally well write $A = \{\{1, 2\}, b, c\}$.

In the previous paragraph all we are doing is defining a set, called $SET2$, whose **individual elements** are the **sets** $\{1, 2\}$, $\{3, 4\}$, and $\{5, 6\}$. In particular, the set $SET2$ has only three elements, because elements are **individual, indivisible** members of a set. For example, $\{1, 2\}$ **itself** is a **set** with two elements, but it is also a **single element** of the set $SET2$.

To continue to get familiar with this idea, let us ask whether or not $SET1 = SET2$. Recall that for these two sets to be equal, we need all of the **elements** in $SET1$ to be in $SET2$, and vice versa.[2] Notice that when we write these two sets down on paper, the only difference between them is the addition of some strategically placed brackets in $SET2$ that are absent in $SET1$. However, it is not the aesthetic appearance of two sets that determine their equality (or lack thereof), but rather their logical content.

By exploring the question of the equality (or lack thereof) of $SET1$ and $SET2$ a bit more, we quickly see that these two sets could hardly be any **less** equal. The reason for this is that **none** of

---

[1] Namely, we recall that two sets are distinct if and only if there is (at least) one element that is in one set but not the other.

[2] The "vice versa" here is extremely important, for if it is the case that two sets are equal then there cannot be an element in one set that is not in the other. A little thought will allow one to convince oneself that this can happen if and only if all of the elements in one set are in the other, **and** that all of the elements in the **other** set are in the first set.

the elements in $SET1$ are in $SET2$, and **none** of the elements in $SET2$ are in $SET1$. For example, the element 1 is in $SET1$, but it is not in $SET2$ because $SET2$ has 3 elements, and none of them are 1. Of course, the element $\{1,2\}$ in $SET2$ is **itself** a set which contains the element 1, but that most certainly does not mean that the element $\{1,2\}$ in $SET2$ is equal to the element 1 in $SET1$. Conversely, we also see that the **element** $\{1,2\}$ is in $SET2$, but not in $SET1$. This is because $SET1$ has 6 elements in it, and they are 1, 2, 3, 4, 5, and 6. None of these 6 elements are equal to the element $\{1,2\}$. The reader is encouraged to convince himself that $SET1$ and $SET2$ do not share a single element.

As a second example of sets of sets, we see that $SET1$ is not the same set as the set $SET3 = \{1, 2, 3, 4, \{5, 6\}\}$. This is because the elements 5 and 6 in $SET1$ are not in $SET3$. Additionally, the element $\{5,6\}$ in $SET3$ is not in $SET1$. The reader is now also encouraged to convince herself why $SET2$ and $SET4 = \{\{1,2,3\}, \{4,5,6\}\}$ are not the same sets.[3]

Power Sets

Let us now continue to explore the idea of "sets of sets" by combining this new idea with an old idea in order to make a useful mathematical definition. As we saw in Chapter 2, any finite set has a certain finite number of distinct subsets. In particular, we saw that if a set has N elements, then it also has 2^N distinct subsets (recall what "2^N" means from Chapter 2). We can now formulate this more precisely by using our notion of a set of sets.

Recall that a subset of a set is itself a perfectly good set. Therefore, since we just saw that a set can be viewed as a perfectly good element of some other set, and since a subset can be viewed as a perfectly good set, we can very naturally define the **set of subsets** of a given set. In other words, if we are given (or if we define) some set and we call it A, then we can immediately also define the set of all **subsets** of A. This set of all subsets of A is precisely what it sounds

---

[3]Hint: Is the element "$\{1,2\}$" (which we know is in $SET2$) in $SET4$? Is the element "$\{1,2,3\}$" (which we know is in $SET4$) in $SET2$? Hint for Hint: No and no.

like — it is the set whose **individual, indivisible** elements consist of the subsets of A. For any set A, we can choose to call this set of subsets anything we like, and the industry-standard name for this set is "the power set" of A. Let us make this a formal definition.

Definition 3.1. For any set A, the set of all subsets of A is called the **power set** of A, and it is usually denoted by $P(A)$.

Note that the symbols "$P(A)$" simply mean "the power set of the set that is within the parentheses." Note also that, using this new terminology, the pattern that we derived in Chapter 2 is simply that if a set A has N elements, then its power set $P(A)$ has 2^N elements (and also that the **elements** of $P(A)$ are the **subsets** of A). It is important to fully appreciate the fact that the elements of A and $P(A)$ are of a completely different type — the elements of the set itself (the set A) are whatever they were to begin with (apples, donkeys, numbers, or whatever), and the elements of the power set (the set $P(A)$) are **subsets** of the set A.

As usual, we will get more comfortable with the definition of power sets by considering an example. Let us define the set $A = \{a, b, c\}$. From the pattern derived in Chapter 2 we know that there should be 8 (which is 2^3) distinct subsets of this set, and indeed there are. Now, however, we have the mental machinery to be able to write down these subsets as elements of the power set $P(A)$ of A. If we denote the empty set[4] by $\emptyset$, then we have the following expression for $P(A)$:

$$P(A) = \{\{a,b,c\}, \{a,b\}, \{a,c\}, \{b,c\}, \{a\}, \{b\}, \{c\}, \emptyset\}.$$

Thus, one **element** of $P(A)$ is $\{a, b\}$, another element is $\{a, b, c\}$, and yet another is $\emptyset$. We note that there are indeed 8 elements in $P(A)$, and we also note that the elements of $P(A)$ are not only sets, but they are even sets with different numbers of elements! Namely, the set $\{a, b, c\}$ is a **single element** of $P(A)$, and the set $\{b\}$ is also a **single element** of $P(A)$. This is simply a result of the fact that we are viewing **entire sets** as **single elements**, and therefore any "internal structure" of the sets as a whole does not matter when we

---

[4] Recall that the empty set is a subset of **every** set.

view the sets as single, indivisible objects (i.e., as elements of some other set).

The power set is just one example of a set of sets, but it is a particularly nice example because it is created "from" another set. In other words, if we are given any set A, then we can always form the set of all of A's subsets. Of course, in the general discussion of sets of sets, we are not limited to only discussing power sets. For example, we can take a set of apples, a set of oranges, and a set of bananas, and then form another set simply by considering each one of our three sets of fruit as a single element of a 3-element set. This set would then definitely be a set of sets (with 3 elements, each element itself being a set of fruit), yet this set would also not be the power set of some other set. There is nothing wrong with considering different kinds of sets of sets, but it is definitely the case that some sets of sets are more interesting than other sets of sets. One gets the feeling that the above set of sets of fruit is much less interesting than, say, the set of all subsets of the positive whole numbers.

This is part of our job as mathematicians — to hone in on those definitions and theorems that are most interesting, and disregard those that are less interesting. We have already seen that mathematics is the art form in which precise definitions are made and their logical consequences are studied. It is a true **art** form, as opposed to an algorithmic procedure, because some of these consequences and constructions are more interesting — and hence more beautiful — than others. It is the artist's job to uncover this beauty. In the next chapter we will see a remarkable conclusion that we have been moving towards — one that will show us that the world of mathematics is more subtle than we could possibly know.

Exercises

1) Let $A = \{1, 2, 3, 4, 5\}$ and let $B = \{\{1, 2\}, \{3, 4\}, 5\}$. How many elements does A have? How many elements does B have? Does $A = B$?

2) Write down (in the bracket notation that we have developed) all of the **elements** in the set $P(A)$ when $A = \{\text{Kobe, LeBron, MJ}\}$.

3) Knowing that a set with N elements has a power set with 2^N elements, how many elements does the power set of the power set have? If we recall that $P(A)$ denotes the power set of A, then the question is to find the number of elements in $P(P(A))$. To start thinking about this problem correctly, first consider the set $A = \{a, b\}$ and write down all of the **elements** of $P(A)$. There should be four such elements. Then $P(P(A))$ is the power set of a 4-element set. How many elements does this set have? Writing them all out explicitly is good practice.

Chapter 4

A Paradox

Here Comes Trouble

In Chapter 1 we studied sets and in Chapter 2 we studied their subsets. In the previous chapter we studied sets whose elements are themselves sets, since after all a set is indeed a "thing," and "things" are the elements of sets. Whether it feels like it or not, we have actually come a long way in these three chapters — so long, in fact, that we almost cannot go any further. Let us see why.

As we may have noticed, everything that we have defined and discussed so far has been in some sense as "obvious" as it can be. We started in Chapter 1 by considering what could be the most fundamental mathematical object, and naturally our answer was "anything." In other words, we can take anything we would like in the whole universe and even those things not in the universe and consider whatever this "thing" is as an element.

We then looked for the second most fundamental concept, and again naturally came to the conclusion that "a collection of things" is the natural next step. After considering one "thing" (an element), what could be more obvious than wanting to consider "some things" (a set)? All of this seems very intuitive, and indeed it is. More importantly, it is extremely general, and therefore extremely powerful. But remember that when it comes to mathematics, intuitiveness means nothing — all that is important is logical consistency.

As it turns out, we have indeed already lost a good deal of logical consistency within just the first three chapters of this volume. This **hopefully** seems surprising, since the whole point of the first three chapters of this book have been to derive mathematics from a logically infallible starting point.

So from where can this supposed logical inconsistency originate? Let us first recall what can be described as "the process of mathematics," which is the procedure of making some definitions and then deriving truths about these definitions using nothing but pure logic. However, we should not have been so cavalier with the idea that we can make "any definition at all," because some definitions are simply not allowed.

How do we know when a mathematical definition is "allowed" or not? This is a very delicate matter, but one that is also supremely important. In order to not get bogged down in the subtleties of these issues, let us take a somewhat ironic point of view and simply **define** a mathematical definition to be one that does not lead to logical inconsistencies. This is motivated solely by our desire to create a framework of mathematics with absolutely no logical inconsistencies. If we limit ourselves to only those definitions that are "allowed" in this sense, as well as to the truths that can be derived from these definitions using only pure logic, then we can be sure that our mathematics is perfectly logical.

Unfortunately, determining the logical consistency of a mathematical definition can be very difficult. This is because **not** proving logical **in**consistency is very different than **proving** logical consistency. Let us think about this for a second. We can **prove** that a definition is **in**consistent by using that definition to arrive at a paradox. For example, if we could start with some definition and prove that $1 \neq 1$, then that definition is surely logically inconsistent since in **any** form of mathematics we want it to be the case that anything always equals itself.

However, suppose we start with some definition and try to prove some kind of paradox from it. Suppose we try for a thousand years (medicine will have to come a long way to let us live that long) yet still fail to arrive at any kind of paradox from this definition. Does this mean that the definition is logically consistent? Absolutely not. It simply means that we have not found any way to show its inconsistency. Of course, it could be the case that the definition that we started with is indeed logically consistent, in which case we have just wasted a thousand years. In most of these volumes (including this one), we will not explore the various methods of proving the logical consistency of definitions, though we will take note of what makes certain definitions "well-defined" at the appropriate times.

In this chapter, we will uncover the fact that the definitions that we have made so far are indeed **in**consistent, and we will derive the paradox that proves this. The reader is encouraged to take a moment to appreciate how significant of a statement this is. We have worked very hard to make sure that all of our statements in the first three chapters of this volume were as logically precise as they could reasonably be. Yet here we are faced with the claim that we have not been careful enough. This shows how amazingly complex and subtle "logic" herself really is. We abide by logic's rules, and not vice versa. In order to satisfy her requirements we must be incredibly careful. In what follows we will uncover where we have gone astray, we will reflect on it briefly, but we will **not** attempt to rectify our mistakes. We will not rectify our mistakes for many reasons, with the primary reason being laziness.

The paradox that we will uncover in this chapter has led to an entire group of people studying "set theory" as an independent field. This field is almost solely aimed at making our notions of sets logically consistent. However, almost all of math can continue to be done **without** worrying about these subtleties, and this is what we will do with the rest of these volumes. So while it is important to be acquainted with these subtleties, one must also see that we can indeed — for all practical[1] purposes — move on with our mathematical adventures and continue our forward progress. So let us begin to see the problem, and then let us immediately forget about it.

---

[1] By "practical" here we do not mean in the technological or economic or societal sense, but rather in the purely logical sense. The idea here is that the logical framework of mathematics **as a whole** is not greatly or immediately threatened by the subtleties that we will soon see.

A Warm-Up

First let us consider the following warm-up paradox. Suppose we are librarians and that we are given two blank books. Suppose also that we are asked to record every book that is in the library into one of these two blank books. In other words, all we need to do is make sure that the title of every book in the library gets written down into one of the two blank books. However we are given some rules as to how this must be done. In particular, we must record these books in such a way that **every** book is recorded **once and only once** in one of the two blank books.

Let us get more specific. Suppose we label our two blank books as "book A" and "book B." We are then told to write down the title of every book in the library into **either** book A **or** book B in such a way that **every** book is recorded **only** once. We cannot miss any books, and we cannot double count any books. There are several ways we could do this. We could write all of the titles down in book A exactly once, leaving book B empty. This does indeed fulfill the requirements! We could also record each book into book A if it has the color red somewhere on its cover, and all the books that have red nowhere on their covers into B. Since every book either has red or does not have red on its cover, we will get each and every book exactly once, and we will therefore fulfill the requirements.

Suppose now that we come up with a different rule for organizing the books. Suppose we are not thinking too clearly and that we therefore come up with the following strange rule: "Record every book that contains its own title somewhere within its text into book A, and record the books that do not contain their own title in their text into book B." Thus, since the phrase "Harry Potter" shows up inside the text of the *Harry Potter* series, all of the *Harry Potter* books get recorded into A. Any book that goes from start to finish without mentioning its own title, however, gets recorded into B. Notice that although this is a somewhat strange rule, it should still satisfy the requirements. Namely, **any** book **either** contains its title within its pages **or** it does not. Therefore, **every** book will be recorded into either A or B. There is no trick up our sleeve here (or anywhere) — this is a perfectly good rule.

However, being the thorough librarians that we are, we realize that book A and book B are also "books in the library." Thus, according to the requirements, books A and B need to be categorized into either book A or book B, just like all the other books in the library. Upon thinking about where to categorize books A and B we become confused, and rightfully so. What book does book B get recorded into? Recall that book B is the book that records all of the books that **do not** have their own title within their pages. We know that book B needs to go **somewhere**, because our rule should apply to **every** book in the library. Let us therefore do what mathematicians do best — guess.

Suppose we record book B into book A, so that we are supposing (i.e., guessing) that book B **does** contain its title within its pages. If book B is recorded in book A, then it is **not** recorded in B since we cannot double count our books. And if it is not recorded in book B, then it does **not** contain its title. This contradicts the fact that it is recorded in book A, since books recorded into book A must contain their own title!

Therefore we are forced to assume that our guess was wrong — book B **cannot** be recorded in book A, like we had guessed. But that means that book B is recorded in book B (since it **must** be recorded somewhere, as our rule should "hit" every book in the library based on whether or not it contains its own title). If book B were to be recorded into book B, it would end up being the case that book B **does** contain its title within its own pages. But this contradicts the fact that, according to our rule, any book recorded into book B does **not** contain its title within its pages! So we cannot even record book B into book B! Book B is simply un-recordable — it belongs neither in book A nor in book B. But this is strange, because the rule of recording books according to whether or not they contained their own title was supposed to work for **every** book, just like we were ordered.

In fact, this is much worse than strange — it is catastrophic. There really is no trick up our sleeve. There is no fancy play on words or hidden meaning. This is not a riddle that needs to be "figured out," where one simply finds the trick in our wording which makes this whole problem go away. What we get is what we see, and what we see is a total breakdown in what seemed to be foolproof logic.

And therein lies the problem! Our logic only **seemed** to be foolproof. In our own minds, every link in the logical chain worked perfectly. We had what seemed to be a perfectly well-defined problem, we came up with what seemed to be a perfectly well-defined rule, we used what seemed to be perfect logic, and we found ourselves led to a perfectly inescapable paradox. This is because at various stages throughout the process, we were not nearly as careful as we needed to be in defining our terms, our rules, our recording process, what rules were "allowed" and which were not, and so on. Our **intuition** — what **seemed** logical — led us astray. Seeing as mathematics does not bend to suit our intuition, we are **forced** to bend our intuition in order to suit mathematics. And in order to do mathematics, we need a system of logic that is free of such paradoxes. We therefore would need to be much more careful about our definitions and rules in order to work our way out of this paradox. But before we discuss[2] how this is done, let us see how this paradox translates into the mathematical language that we have developed so far, and uncover the "actual" (i.e., non-warm-up) paradox that this chapter is designed to show.

The Main Event

We will now reformulate the paradox from the previous section in the more precise language of sets and subsets. We know that we can discuss sets of sets, where entire sets are themselves the **individual** elements of some set. Moreover, we know that a set is just a collection of elements that can be distinguished from each other. We can define the set of all apples, or the set of all basketballs, or the set of all whole numbers, since in each of these cases we know how to distinguish the elements in the set from each other. In other words, once we have a collection of elements that we can distinguish from each other (which is a pretty easy requirement to fulfill), then we can form a set out of

---

[2]Recall that we will not **really** discuss how this is done, as there is an entire industry of mathematicians — who have created a remarkably large literature — exploring various avenues of this problem, and we would simply get bogged down if we were to enter into it all.

those elements. Accordingly, since we know how to distinguish any two sets — i.e. we can determine whether any pair of sets are equal or not, according to whether or not they have the same elements — we can therefore make up arbitrary collections of sets. We can define the set of all sets of numbers,[3] or we could define the set of all finite sets of basketballs. Our most significant example, however, will be **the set of all sets**. That's right. We can, in principle, define the set of **all** sets — so far there is nothing in our formalism that forbids us from doing so, since we know how to distinguish one set from another and therefore the elements of this set are perfectly well-defined.

We are now forced to realize that the set of all sets contains **itself** as an element. After all, this set is the set that contains **all** sets as elements. Since it is a set itself, it must contain itself as an element. Thus, **the set of all sets contains itself**.[4] Since we have found one set that contains itself as an element, we now know that the statement "some sets contain themselves as elements" is true. Of course, **most** sets do **not** contain themselves as elements. For example, the set {1, 2, 3, 4} does not contain itself as an element because it only contains the elements 1, 2, 3, and 4, whereas the set **itself** is a set of four numbers! Note also that the set {a} does not contain itself either. This is because it contains only the element a, whereas the set **itself** is the **set** made up of only the element a. In other words, the set {a} contains the element a, but it does **not** contain {a} itself as an element.[5] Thus we see that we have to very carefully distinguish between a set and its elements, but this is precisely what we have always done — a set of 10 apples does not contain itself because "a set of 10 apples" is not an **element** of this set (the elements of "a set of 10 apples" are apples, not sets of apples).

We now know that some (indeed, most) sets do not contain themselves as elements, and that some (at least one) sets do contain themselves as elements. Therefore it may be interesting to study "the set

---

[3]Note that the elements of this set would **not** be numbers, but rather the elements of this set would be **sets** of numbers.

[4]For the curious reader, there goes the solution to the final exercise in Chapter 1 — now we know why we said it is not obvious!

[5]To illustrate this point further, the set $\{1, \text{apple}, 37, \text{Kobe}, \{a\}\}$ does contain $\{a\}$ as an element, and the set $\{a, b, c\}$ does not.

of all sets that contain themselves" as well as "the set of all sets that do not contain themselves." In other words, if the enemy hands us a set, we can ask if it contains itself as an element or not. Once we figure out the answer to this question, we place this set (the set that the enemy handed us) in "the set of all sets that contain themselves" if it contains itself, and we place it in "the set of all sets that do not contain themselves" if it does not. Notice that we should be able to place **every** set in **either** "the set of all sets that contain themselves" or "the set of all sets that do not contain themselves," since every set either contains itself or it does not (this is the same as saying that every book in the library is either red or not red, or that every book in the library either contains its title within itself or it does not).

According to the logic of the previous paragraph, **every set imaginable** should be an element of either "the set of all sets that contain themselves" or "the set of all sets that do not contain themselves," since every set imaginable must either contain itself or not.[6] Therefore, "the set of all sets that do not contain themselves" should be in one of these two sets since it is, after all, a set. Let us see which set it belongs to by first making a guess. If "the set of all sets that do not contain themselves" is in "the set of all sets that do contain themselves," then "the set of all sets that do not contain themselves" would both contain itself (because it would be an **element** of "the set of all sets that contain themselves") and not contain itself (because it would also **not** be an **element** of itself, i.e., "the set of all sets that do not contain themselves," since it **is** an **element** of "the set of all sets that **do** contain themselves"). The fact that a particular set simultaneously contains and does not contain itself is clearly a contradiction.

Therefore, in order to do away with the contradiction that we just arrived at, "the set of all sets that do not contain themselves" should be in "the set of all sets that do not contain themselves." This is because we know that "the set of all sets that do not contain themselves" has to be in one or the other of these two sets (and we just showed that it cannot be in "the other"). But then "the set of

---

[6]Note that "the set of all sets that contain themselves" and "the set of all sets that do not contain themselves" are playing the same role here that the books A and B did in the library paradox of the previous section.

all sets that do not contain themselves" actually does contain itself, and therefore should **not** be an element of "the set of all sets that do not contain themselves"! We therefore have another contradiction! We have found a set that cannot be categorized into either of these two sets, despite the fact that these two sets ("the set of all sets that do contain themselves" and "the set of all sets that do not contain themselves") were supposed to contain, between the two of them, every imaginable set. As in the library paradox, there truly is no trick up our sleeve. This is a genuine — and fundamental — flaw in how we have been manipulating sets. Everything that we have done so far (just as in the library paradox) has seemed perfectly reasonable, justified, and rigorous. Nevertheless, we have arrived at a full-fledged paradox: the existence of a set that cannot be placed in either of two sets that are meant to, between the two of them, contain **all** sets.

In order to avoid contradictions like these, we have to be a lot more careful about how we define and manipulate sets, despite the fact that what we were doing before seemed so obvious and intuitive. Again, it is not up to us to decide how mathematics should behave, nor to hope that mathematics should be in line with what we think is obvious or intuitive. Rather, it is up to us to align our logic and intuition with its behavior.

The details of how we should describe and manipulate sets take us far away from the type of math that we would like to explore here, so we will leave the topic of fixing the above paradox alone for now. We will eventually find a way out of these problems using what is called Category Theory, but that will not be for a while. The truth is that a good deal of mathematics can be discovered by using the intuitive description of sets that we established prior to this chapter, and so that is what we will continue to do. Note that this most certainly does **not** mean that the mathematics that we will discover[7] together in these volumes is in any way placed on unsturdy foundations. The math that we will describe will hold even with the more careful, detailed, and technical definitions of sets. All we are doing is avoiding that tedium, but we are not "dumbing down"

---

[7]Technically, "rediscover," since all of the math that we introduce in these volumes has already been very well-established by many others over the course of a very long time.

mathematics, or deriving untrue statements. So long as we do not deal with sets that lead to contradictions like the one above (which is known as Russell's paradox), we will be deriving unassailably true and eternal statements.

To recap, the whole point of this chapter was to show that even in the seemingly simple theory of sets, there is immense subtlety and beauty (if the reader does not find logical subtlety like this beautiful, she should not lose heart — there will be more beautiful things to come). For now, we happily admit the existence of these subtleties, but we leave them behind and bulldoze our way into cooler and cooler math...

Chapter 5

Functions

Getting Some Intuition

Now that we know all about sets (enough, in fact, to derive the paradox of Chapter 4), there is a very natural question to ask: What happens when we have two of them? In other words, what kinds of things can we learn when we have not one, but two sets around? In particular, what types of relationships can exist **between** sets?

We will formalize — i.e., make precise — this question soon, but let us first comment on the generality of this line of questioning. Perhaps the most pervasive idea studied in mathematics is the concept of "relationships between structures." A mathematician first defines some new object (in our case the object that we have defined is a set) and then studies the behavior of "relationships" between two such objects. As we will see, there is always a remarkably subtle interaction between the mathematical object itself, and the relationships that can be established between two such objects. One of the most general types of mathematical object is the set, and therefore the relationships that can be established between sets are some of the most general as well. Let us therefore quit the philosophizing for now and move on to studying these relationships.

The most basic and fundamental relationship between two sets is known as a **function**. We will make a somewhat precise definition shortly, but first let us describe the intuitive idea behind what a function does.[1] Intuitively, the purpose of a function will be to go "from" one set "to" another set[2] in the sense that it "sends" every element **from** one set **to** some element of the other. Of course, "out there" in the world of mathematics there is nothing that literally picks up elements in one set and carries them over to some other set, and this is simply because as far as we know there is no physical location in which sets and functions actually exist in some tangible sense. We therefore need to formalize this idea as much as we can.

When we say that a function "sends" an element in one set to an element in another set, we really mean that it "associates" the former element to the latter. One possible physical analogy to keep in mind is the following. Suppose we have a set of apples and a set of baskets, and suppose we want a function **from** the apples **to** the baskets. It then must be the case (according to how we will end up defining functions) that **every** apple must be associated to **some** basket. Let us say that "association" means "putting the apple in the basket." There are many possible such functions. We could put all of the apples into the same basket, or we could (assuming there are enough baskets) put one apple in each basket. We could put 10 apples in one basket and the rest of the apples in some other basket. And in all of these possibilities, we have tremendous freedom in deciding which apples go to which basket(s), meaning that there are tremendously many different functions from our apples to our baskets.

If, however, we want a function from a set of dogs to a set of NBA players, then there is no natural analog to the idea of putting the apple "in" the basket. This is why our mental picture of a function is **not** that of physically moving an element in one set to an element in another, but rather just that the function "associates," or "assigns"

---

[1] When studying math it is always important to gain **intuition** for — i.e., a good mental picture of — any mathematical object or theorem before attempting to understanding the precise, rigorous definition or statement.

[2] Note that these two sets (the "from" set and the "to" set) are completely arbitrary sets. This simply means that we do not necessarily need to specify that one set is, say, a set of basketballs and another set is a set of numbers. Functions can exist between **any** two sets, and so when we define functions we will leave the two sets that the function "goes between" completely arbitrary. It may even be the case, in some circumstances, that the two sets (the "from" set and the "to" set) are the same!

the two elements to each other. One good picture to keep in mind for this is that a function assigns an arrow **from** each element in one set **to** an element in the other set. Thus, instead of physically putting the apples in the baskets, let us have our function simply place an arrow next to each apple, pointing to whichever basket that apple is associated with.

Let us emphasize again that a function not only assigns elements of two sets to each other, but that it also must "go" from one set to another.[3] Whichever set the function "starts at" must have every one of its elements assigned to some element of the set it is going to. We also note that absolutely none of this talk about functions is in any way "written in the stars." There is no cave in which the first human found a book that told her what a function is. Instead, we are simply **defining** a function to be a way of assigning elements in one set to elements in another, and we are also defining functions so that **each** element in the "from" set goes somewhere in the "to" set (i.e., no element in the "from" set is skipped). We could have made many other definitions, but this definition turns out to be quite useful. The only things that **are** "written in the stars" are the **consequences** that we can derive from this definition of what a function is. Let us see an example to make all of this concrete.

Consider the set $A = \{1, 2, 3\}$ and the set $B = \{a, b\}$. A function from A to B is an assignment of every element in A to some element in B. We could assign[4] $1 \to a, 2 \to a$, and $3 \to b$, and this would be a perfectly good function because we have given a "value" to every element in A. Quotes are again used here on "value" because the "value" of an element in A is some element in B. For example, the "value" of 1 under this function is a, and the "value" of 2 under this function is a as well. Normally we like to think of "values" as being numerical, but we really can generalize the notion of "value" to any set. If A is given as above and $C = \{$donkey, horse, cow, chicken$\}$,

---

[3]There are quotes here because we are emphasizing that there is not really anything moving "out there" in the world of math, and so the word "go" is not exactly correct. Nevertheless, the dynamical picture of a function in terms of things actually moving is a good analogy to keep in one's head.

[4]In what follows, we usually use the terminology "1 goes to a, 2 goes to a, and 3 goes to b."

and if we define the function 1 → horse, 2 → chicken, 3 → cow, then the "value" of 1 is horse, and the "value" of 3 is cow.

There are a couple of things that we need to note about these examples that will in fact be crucial for when we actually define a function rigorously. The first thing to note is that although every element in A (the "from" set) is given a value in C (the "to" set), it is not the case that every element in C has some element in A assigned to it. For example, the element donkey in C never got an element in A assigned to it. That is fine, because the function is defined to be **from** A and **to** C, and not the other way around. We want to define a function so that **every** element in the "from" set has something in the "to" set assigned to it, but this does **not** require us to use **every** element in the "to" set when doing so. We could, for example, assign "chicken" to every element in A, and that would be a perfectly fine function. I.e., the function "1 → chicken, 2 → chicken, 3 → chicken" is perfectly fine because nowhere in our definition[5] of a function did we exclude the possibility that we could assign the same value to every element in the "from" set.

The last thing to note is that we are only assigning one element (i.e., one "value") to every element in A. In other words, the function "1 → cow, 2 → chicken, 3 → cow, 3 → donkey" is not a function because 3 is sent to both cow and donkey, which is not okay. This is not okay because we simply **define** — i.e. force — a function to be something which assigns **only one** value to each element in the "from" set. This is analogous to the fact that we cannot put one apple into two different baskets (without cutting the apple, of course).

Let us emphasize once again that we could, of course, have defined a function to be otherwise. We could have defined a function to be something that always assigns **two** elements to each element in its "from" set. We could also have defined a function so that it did **not** have to assign a value to **every** element in its "from" set. We are simply choosing to define a function in the way that we have because it appears to be one of the most interesting definitions to study. Remember, once we define some object (in this case, a function), the logical deductions that can be made about that object are completely

---

[5]Note that we have only talked about the definition of a function, and that the precise definition will come further below.

out of our control (or "written in the stars" as mentioned above). We **can** choose the definition, but we **cannot** choose the deductions because the deductions are firmly rooted in the world of mathematics, which is already "out there," beyond us. It just so happens that this definition of a function is much more interesting than other definitions, which means that we have stumbled across the definition that is somehow "right." Let us therefore just go ahead and make this definition and then we will study its consequences in later chapters.

Definition 5.1. Let A and B be two arbitrary sets. A **function** from A to B is an assignment to every element in A exactly one element in B.

That is it. There are no requirements for how many elements of B we end up using overall, and there are no requirements for which elements go where. Thus, this is an extremely general definition, since all we need to remember is that every element of A needs to go to **some** element in B, and that each element in A **only goes to one** element in B.

To close out this section, let us introduce a nice way to visualize functions. Suppose we have the sets $A = \{1, 2, 3\}$ and $B = \{a, b\}$, for example. Suppose also that we have a function f that goes from A to B, and that f is defined by the relations $1 \to b$, $2 \to b$, and $3 \to a$. In other words, we are **defining** f to be the function from A to B that sends 1 (in A) to b (in B), 2 (in A) to b (in B), and 3 (in A) to a (in B). We can visualize this by drawing an arrow **from** the elements in A to the elements in B that they get sent to. This function is then represented by the arrows drawn in Figure 5.1. All we are doing here is looking at each element in A, seeing where f sends that element, and drawing an arrow that starts at the element in A and points to the element in B to which it is sent. This is a very useful visualization tool, and we will get some more practice with it in the exercises at the end of this chapter.

NOTATION AND TERMINOLOGY 53

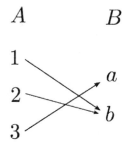

Figure 5.1: A visual representation of the function defined above.

Notation And Terminology

Let us now set up some notation.[6] Suppose we have two sets A and B and a function from A to B. Let us denote this function by f. In particular, f is a noun — it is a function.[7] Let us now agree to denote the entire phrase "f is a function from A to B" simply by the collection of symbols

$$f : A \to B.$$

Thus, in the small amount of writing "$f : A \to B$," we know that A and B are sets, we know that f is a function between them, and we know that f goes **from** A and **to** B (i.e., that every element in A has one and only one element in B associated with it via f). We sometimes also call A the **domain** of f, and B the **codomain** of f. In this sense, then, A is the "domain of influence" of the function f (since f sends every element of A somewhere), and the word codomain is used just to remember "where f sends things to." Accordingly, if g were a function from a set C to a set D, we would write $g : C \to D$. Here, C would be the domain, and D would be the codomain. Let us summarize this with the following "two in one" definition:

Definition 5.2. Let f be a function from A to B, so that $f : A \to B$.

---

[6]Remember, notation is not scary, we just need to learn what it is.

[7]Try not to think of f as being the first letter of the word "function," because we could have just as well called this function from A to B "g," or "Donkey," or "Kobe." The letter f is simply an arbitrary name that we are giving this function.

Then A is defined to be the **domain** of f, and B is defined to be the **codomain** of f.

The last bit of notation that we will set up is a succinct way to describe where a function f sends individual elements. In other words, we will set up notation that not only encapsulates where a function starts and finishes (i.e. f starts at A and finishes at B), but also where the individual elements go. Suppose $A = \{1, 2, 3\}$, $B = \{\text{cat, donkey, apple}\}$, and $f : A \to B$ (read aloud as "f is a function from A to B," so that this sentence is in fact grammatically correct). Moreover, suppose f is the function which assigns $1 \to \text{cat}$, $2 \to \text{apple}$, and $3 \to \text{apple}$. Then we simply denote this by $f(1) = \text{cat}$, $f(2) = \text{apple}$, and $f(3) = \text{apple}$. I.e., the notation $f(1) = \text{cat}$ is simply shorthand notation for the phrase "f is a function that sends 1 to cat." We will get more practice with this notation in the exercises of this chapter as well as in later chapters.

Just as sets may have seemed trivial until we were able to derive a highly non-trivial result from them (namely, the paradox of the previous chapter), functions may too at first sight seem like a rather trivial concept. Nonetheless, they form a cornerstone of mathematics and are absolutely pivotal in understanding virtually all of the field. In fact, in only a few chapters we will use functions to make precise the notion that there are, in fact, some infinities which are in a clear sense "more infinite" than others. For now, let us be patient and simply internalize the idea of a function, and the corresponding vocabulary and notation. To do this, we turn to the exercises.

Exercises

1) Let $A = \{a, b, c\}$ and $W = \{14, 57, 85, 0\}$. State whether or not each of the following are functions from A to W.

   (a) The assignment that is defined by $f(a) = 57$ and $f(b) = 0$.

   (b) The assignment that is defined by $f(a) = b$, $f(b) = 85$, and $f(c) = 14$.

   (c) The assignment that is defined by $f(a) = 57$, $f(b) = 57$, and $f(c) = 85$.

   (d) The assignment that is defined by $f(a) = 0$, $f(b) = 57$, $f(a) = 14$, and $f(c) = 57$.

2) Let $\text{Num}_1 = \{1, 2, 3\}$ and $\text{Num}_2 = \{4, 5, 6\}$. Define three different functions from Num_1 to Num_2. Namely, come up with a function f such that $f : \text{Num}_1 \to \text{Num}_2$, and then do it two more times (coming up with different functions). Then define three different functions from Num_2 to Num_1. For each function that you define, consider drawing the corresponding "arrow diagram," analogous to Figure 5.1.

3) Let $C = \{\text{Mt. Everest}, \text{cheeseburger}, 4\}$. Define two different functions from C to itself. Namely, come up with a function f such that $f : C \to C$, and then do it again by defining a different function. Again, consider drawing the "arrow diagrams" for the functions you define.

4) (a) Let $A = \{a, b\}$ and let $X = \{0, 1\}$. How many possible functions are there from A to X?

   (b) Let $A = \{a, b, c\}$ and let $X = \{0, 1\}$. How many possible functions are there from A to X?

Chapter 6

Injective, Surjective, Bijective

Adding Structure

Before we panic about the scariness of the three words that title this chapter, let us remember that terminology is nothing to be scared of — all it means is that we have something new to learn! As we will see by the end of this chapter, these three words are in fact not scary at all. Mathematicians have a funny way of assigning very fancy words to ideas that are not very deep,[1] and this is one such instance. Let us therefore go ahead and just learn what these words mean, and in the following chapters we will see why these words (and the ideas that they represent) are useful.

In the previous chapter we learned about functions between sets, thus giving us a mechanism for relating elements in one set to elements in another set.[2] Let us recall that a function is an incredibly general object — all that is needed is two sets and a relation between them such that one of the two sets has every one of its elements

---

[1] Of course, sometimes the fanciness of the words involved does reflect the depth of the idea behind those words, but not always.

[2] Keep in mind, though, that this "other" set could be the set itself. For example, if $A = \{1, 2, 3\}$, then a function f from A to A such that $f(1) = 2$, $f(2) = 1$, and $f(3) = 3$ is a perfectly good function, as is a function g from A to A such that $g(1) = 1$, $g(2) = 2$, and $g(3) = 3$.

associated with exactly one element in the other set.

And that is it. There were no other requirements for a particular relationship between sets to be a function. In particular, if $f : A \to B$ (recall that this reads "if f is a function from the set A to the set B"), then f could send every element in A to the same element in B, or it could send elements in A to lots of different elements in B, or anywhere in between. There also were not any requirements on how many elements in B needed to be "hit" by the function.[3] A function is simply a rule that assigns to each element in A exactly one element of B. A given function might have some additional special properties about it — for example, it **may** very well be the case that every element in B is indeed hit by some element in A — but any additional property that the function has is just a bonus.

The generality of functions comes at a price, however. The cost is that it is very difficult to prove things about a general function, simply because its generality means that we have very little structure to work with. Thus, what we want to do is focus on certain kinds of functions. I.e., we want to limit ourselves to considering functions that have specific properties so that we can use these new properties to prove things. This is in fact a very general pattern in mathematics: we define some very general object (a set, a function, or whatever else) and then slowly start to "add structure" to it so that we can prove things and ask new questions. By "add structure" we simply mean that we give certain extra properties to the object so that even though it loses some of its generality, it gains a certain structure that allows us to ask more interesting questions about it.

Injectivity

What type of "additional structure" do we want to consider here with functions? Let us consider again what a function is, and ask what the most obvious kinds of properties we would like some func-

---

[3] Here, when we say that an element in B is "hit" by an element in A via the function f, we simply mean that the element in B has **some** element in A that gets mapped to it. We use the word "hit" because when an element in B is "hit" by an element in A (via f), the end of the arrow lands on the element in B and so we can say that the arrow "hits" this element in B.

tions to have. For the rest of this chapter, let us let $f : A \to B$ (note again that this is read as "let us let f be a function from the set A to the set B" so that the sentence is indeed grammatically correct[4]), where A and B are two completely arbitrary sets — i.e., we do not need to specify the concrete elements of A or B, and can instead say something about f **regardless** of what A and B are.

First, recall that f can send two different elements in A to the same element in B. For example, if $A = \{$cat, dog, horse$\}$ and $B = \{$cheese, crackers, wine$\}$, then the function defined by "f(cat) = cheese, f(dog) = cheese, and f(horse) = wine" is a perfectly good function, despite the fact that cat and dog are both sent to cheese. Suppose, however, that f were a function that does **not** have this property for any elements in A. Namely, suppose that f does **not** send any two distinct elements in A to the same element of B. If this is true about f, then we call this function **injective**.[5] Let us see an example.

Let A be a set of boys and B be a set of girls, and let f be the function of "a school dance." Namely, let f be a function that assigns the boys in A to dance with the girls in B. If f were just any general function, then it could be the case that all the boys are dancing with the same girl (surely an uncomfortable experience), or it could be the case that five boys are dancing with one girl, and the rest of the boys are dancing with some other girl (still rather uncomfortable). Indeed, **any** assignment of boys to girls such that **each** boy was dancing with **some** girl — and such that **each** boy was dancing with **only** one girl — would be a perfectly good function **from** the boys **to** the girls. Note that these requirements still allow for it to be the case that more than one boy dances with the same girl, and for it to be the case that some girls do not have any dancing partners.[6]

---

[4]It is important that the mathematics that we write down is grammatically correct once the necessary substitutions of words for symbols is made.

[5]This is simply a definition. We could have chosen to call this type of special function anything we want, but for various reasons the mathematical community has agreed to call this type of function an injective function.

[6]Note that having this function go **from** the boys **to** the girls was an arbitrary choice. Therefore we could have very well defined the function to go **from** the girls and **to** the boys (so that the function f would be $f : B \to A$), in which case it may be the case that more than one girl is dancing with the same boy, and/or

INJECTIVITY 59

Assuming that there is an appreciable number of boys in set A and girls in set B, there will be a **gigantic** number of possible ways to define a function from the boys to the girls. Of this huge number of functions, there will be a much smaller number of functions for which it is the case that each boy was assigned to a different girl. In other words, for certain special functions from the boys to the girls, no girl will be dancing with more than one boy. Let us agree (amongst ourselves and with the entire mathematical community) to call functions of this kind "injective functions."

There is an important quality about injective functions that becomes apparent in this example of the school dance, and which is important for us in defining an injective function rigorously. Suppose we were told that (i.e., suppose we were **given** the fact that) our "school dance function" is injective. Suppose also that we were told that "boy 1" was dancing with "girl 17," and that "boy 56" was also dancing with "girl 17." Then we would immediately know that boy 1 and boy 56 are actually the same person! Namely, we would immediately know that whoever labeled the boys must have given the label "boy 1" and "boy 56" to the same boy. This is because we know that no two different boys are dancing with the same girl (because we were **guaranteed** that the function is injective). We can then quickly expand on this logic and see that the quality of our "school dance function" being injective is **precisely** the same as knowing that if "boy a" and "boy b" are dancing with the same girl, then "boy a" and "boy b" are actually the same person.

This logic then readily generalizes to the case of not just a set of boys and a set of girls, but also to the case of any two sets at all. Thus, an injective function f from a set A to a set B is **defined** to be one such that if a and b are both elements[7] in A, and if f sends

---

that some boys have no dancing partner. However, we just happened to make the choice that we made and so we will stick with it.

[7]There is a **very** important distinction to make here, as it is often the cause of great confusion. Indeed, it is absolutely impossible to overstate the importance of understanding this footnote deeply, for without a clear understanding of what we are about to explain, this whole series (and all of math) will forever remain excruciatingly confusing. We will often use symbols like a or b to refer to **general** elements in a set, and this should be distinguished from the cases when the letters a and b are themselves **particular** elements in a set. Namely, we might say

a and b to the same element in B (i.e., if $f(a) = f(b)$), then a and b must be the same element in A (i.e., $a = b$)! Let us therefore make this a definition:

Definition 6.1. Let $f : A \to B$ be a function from the set A to the set B. Then f is **injective** if for any elements a and b in A, $f(a) = f(b)$ implies that $a = b$.

Said less formally, this definition tells us that a function is defined to be injective if any time two elements in A are sent to the same element in B, they must in fact have been the same elements in A to begin with! Namely, if $f : A \to B$ is injective, then by definition we know that no two **distinct** elements in A get sent to the **same** element in B.

Let us end this section by looking at an example of an injective function and its corresponding arrow diagram. Let $A = \{1, 2, 3\}$ and $B = \{a, b, c, d\}$. Let us then define the function $f : A \to B$ by the assignments $f(1) = b$, $f(2) = a$, and $f(3) = d$. We note that $f : A \to B$ is indeed an injective function, since no two elements in A go to the same element in B. The corresponding arrow diagram of this f is then as shown in Figure 6.1.

From Figure 6.1 — as well as upon reflection on the definition of injectivity — we see that an equivalent way of saying that a function $f : A \to B$ is injective is to say that every element in B has **at most** one arrow pointing to it.

---

something like "if a is an element of A, then a has such and such property," and the meaning of this statement is that **any** element in A has such and such property. For example, if we let A be the set of positive even numbers, then we can say "if a is an element of A, then a is a positive even number." This is very different from the role of a if we consider the set $\{a, b, c\}$, for example, because now the letter a itself is a **particular** element of our set. In any given case the context of our discussion should make clear whether a (or b, or some other symbol) is a particular element in our set or if it is simply a bookmark for any element in our set. The important thing to note when it is the case that a (or b, or some other symbol) is playing a "bookmark" role is that we could indeed use **any** symbol at all. For example, in the case when A is the set of positive even numbers, we can also say "for any element $DONKEY$ in the set A, $DONKEY$ is a positive even number." In our current case above, the symbols a and b are playing the role of being bookmarks for any general elements.

SURJECTIVITY AND BIJECTIVITY

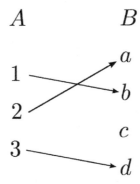

Figure 6.1: The arrow diagram for the function defined above.

Surjectivity And Bijectivity

Injectivity of a function is not the only kind of "specialness" (i.e., added structure) that a function can have, and so we now turn our attention towards a new type of added structure for a function.

Let us again consider the school dance and the function assigning boys to girls. Even if the function is injective, it is not necessarily the case that every girl has a boy to dance with. Namely, there might just be more girls than boys. In this case, even if only one boy is assigned to dance with any given girl, there would still be girls left out. The situation is of course worse[8] if several boys are allowed to dance with the same girl (i.e., if the function is not injective).

Suppose, however, that there are enough boys for each girl to have a dancing partner. This still does not **necessarily** mean that each girl will have a dancing partner because we could, for example, assign all the boys to dance with the same girl (leaving all the rest partnerless), while still satisfying the requirements of a general function. If, however, we assigned the boys in such a way that **every** girl did indeed have a dance partner (perhaps more than one, if the function is not also injective), then the function is called a "surjective function."

---

[8]More precisely, there would be even more girls without dancing partners — whether or not this is truly a **worse** situation is dependent on the reader's point of view.

Notice that surjectivity says nothing about how many boys are dancing with a certain girl, but rather only that each girl has **at least** one boy dancing with her. If there were 10 boys (labeled "boy 1," "boy 2," and so on), and 4 girls (similarly labeled), then we could assign boys 1 through 7 to dance with girl 1, and then boy 8 to dance with girl 2, boy 9 with girl 3, and boy 10 with girl 4. Then each girl has a partner despite the fact that they have very different numbers of them. The way that we want to phrase this in order to later make a rigorous definition is as follows. If our "school dance function" is surjective, then it is the case that for **any** girl, we can find some boy that gets assigned to dance with her. It may be the case that more than one boy is assigned to dance with some given girl, but the important point is that for any girl, **some** boy is assigned to dance with her. Notice that this is nothing but a different way of saying "every girl has at least one boy to dance with."

We can again generalize this idea from the case of a set of boys and a set of girls at a school dance to the case of **any** two sets A and B, and **any** function f from A to B. Namely, we will call f a surjective function if it is the case that for any element b in the set B, there is **some** element a in the set A that gets assigned to b by the function f (i.e., there is some element a in A such that $f(a) = b$). There may be more than one such element in A for any given b in B, but the important thing is that **at least** one such element exists. We are now ready to make the following definition:

Definition 6.2. Let $f : A \to B$. Then f is said to be **surjective** if for every element b in B, there is some element a in A such that $f(a) = b$.

Again, this just means that we **define** a function to be surjective if it happens to be the case that for every element in its codomain[9] we can go find some element in its domain that gets sent to it. It might be the case that there are several different elements in its domain that we could go find, but as long as we can always find one, then we call the function surjective. We have in fact already seen an example of a surjective function, and this is from the previous chapter when we defined the function depicted in Figure 5.1. From that figure — as

---

[9]Recall the definition of "codomain" from the previous chapter.

well as upon reflection on the definition of surjectivity — we see that an equivalent way of saying that a function $f : A \to B$ is surjective is to say that every element in B has **at least** one arrow pointing to it.

We end this chapter with a definition that we will end up exploring a lot more in the coming chapters. Given what we have defined so far in this chapter, it is in fact a very obvious definition to make.

Definition 6.3. Let $f : A \to B$. Then f is defined to be **bijective** if it is both injective and surjective.

Note that this definition is indeed meaningful. Namely, we have seen that we can have functions that are injective and not surjective,[10] and we can have functions that are surjective but not injective.[11] Thus, we are further limiting ourselves by considering bijective functions. I.e., the class of bijective functions is "smaller" than the class of injective functions, and it is also smaller than the class of surjective ones. Moreover, the class of injective functions and the class of surjective functions are each smaller than the class of all generic functions. One way of visualizing this is given in Figure 6.2.

Let us take a moment to consider what would happen if we were to draw the arrow diagram of a bijective function $f : A \to B$. Since a bijective function is, by definition, an injective function as well, we know that every element in B can have **at most** one arrow pointing to it. However, since a bijective function is also a surjective function (by definition), we also know that every element in B must have **at least** one arrow pointing to it. The only way to satisfy both of these requirements at the same time is to have it be the case that every element in B has **exactly** one arrow pointing to it. We therefore note that **there are no bijective functions** from the set $A = \{1, 2, 3\}$ to

---

[10] For example, consider the school dance. If there are more girls than boys, then we can assign boys to girls in such a way that each girl only has **at most** one dance partner (so that the function is injective) but we cannot assign a dance partner to each girl (so that the function is **not** surjective).

[11] If there are more boys than girls, then we can assign a dance partner to each girl (so that the function is surjective), but since we must send each boy to **some** girl (since this is required for our assignment to actually be a function), we **must** send more than one boy to at least one of the girls (so that our function is **not** injective).

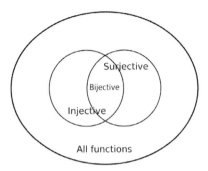

Figure 6.2: A logical picture of the relationship between general, injective, surjective, and bijective functions. Note that the injective and surjective portions **include** the bijective portions, since a bijective function is both injective and surjective, but an injective or surjective function is not necessarily bijective.

the set $B = \{a, b, c, d\}$. This is because there are not enough elements in A to "hit" every element in B, therefore we must miss at least one element in B, and therefore no function from A to B can be surjective. Similarly, we note that **there are no bijective functions** from the set $A = \{1, 2, 3\}$ to the set $B = \{a, b\}$. This is because there are too many elements in A — once we send 1 somewhere, we must send 2 somewhere different, and then we must send 3 somewhere different yet again. But there are not enough choices in B to allow for this, and thus no function from A to B can be injective. There are, however, bijective functions from $A = \{1, 2, 3\}$ to $B = \{a, b, c\}$. For example, the assignments $f(1) = b$, $f(2) = c$, and $f(3) = a$ is bijective, and its arrow diagram is shown in Figure 6.3.

We note that indeed, every element in B in Figure 6.3 has **exactly** one arrow pointing to it. Let us now close this chapter with some remarks and then continue to get practice with functions in the exercises.

In this chapter we have lost some generality by introducing the notions of injective, surjective, and/or bijective functions, since each of these three classes of functions is "smaller" than the class of generic functions. Namely, functions that are either injective or surjective (or

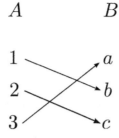

Figure 6.3: The arrow diagram of a bijective function.

both, i.e., bijective) are themselves generic functions as well, but with **additional requirements** that they satisfy. In losing this generality, we have gained the ability to describe more detailed structures within the class of general functions. The concepts of injectivity, surjectivity, and bijectivity will prove very fruitful in the coming chapters, where we introduce a new, more abstract, and more correct method of counting, which will help us work our way towards seeing that there is more than one type of infinity. For now, though, let us get more comfortable with these ideas by looking at the following exercises.

Exercises

1) Define the set $A = \{\text{cow}, 4\}$ and the set $B = \{\text{Magic}, \text{Bird}, \text{Mamba}\}$. Are there any injective functions from A to B? Are there any surjective functions from A to B? Are there any bijective functions from A to B? For each question that has a positive answer, find out how many there are. I.e., if there are injective functions from A to B, find out how many different possible injective functions there are (this is a hard, but possible question to answer — see the solutions after giving it some thought).

2) Define two different surjective functions from the set $A = \{1, 2, a, b\}$ to the set $B = \{\text{cow}, \text{horse}\}$.

3) How many different bijective functions are there from the set $A = \{1, 2, 3\}$ to the set $B = \{a, b, c\}$?

Chapter 7

Infinity And Hotels

Dealing With Infinite Sets

Most of the sets that we have explicitly written down thus far in this volume have been finite sets — i.e., sets with finitely many elements. For example, we have dealt with the sets $\{1, 2, 3\}$ and $\{a, b\}$ quite a lot, and these two sets have three and two elements, respectively. It is very easy to deal with sets when they only have a few elements in them. In particular, if we want to define a function from $A = \{1, 2, 3\}$ to $B = \{a, b\}$, then we can do so by simply and explicitly stating where each element in A goes in B. For example, we could define the function $f : A \to B$ by giving the assignments $f(1) = a$, $f(2) = a$, and $f(3) = b$.

However, there is no reason why a set cannot have infinitely many elements. Indeed, almost all of the important sets in mathematics happen to contain infinitely many elements. One very important set is the set of all positive whole numbers, which we can simply denote by $\{1, 2, 3, ...\}$ where the "..." simply means that the pattern established prior to the "..." will continue on forever. In particular, the pattern here is just increasing whole numbers by one, starting at the number 1, and so this set is defined by the (infinitely many) elements obtained by increasing whole numbers by one forever and ever. This set has infinitely many elements, and we usually denote the set by ℕ and call[1] the elements in the set "natural numbers." Thus we have $\mathbb{N} = \{1, 2, 3, 4, 5, ...\}$.

The set ℕ is a perfectly fine set to define, but we are immediately faced with a difficulty. Namely, how are we able to define a function from the set ℕ to, for example, the set $B = \{a, b\}$? If we called our function $f : \mathbb{N} \to B$, then once we started writing down the explicit assignments we would see that we are in for a world of hurt. For suppose we said that $f(1) = a$, and that $f(2) = b$, and that $f(3) = a$, and that $f(4) = a$, and that $f(5) = b$, and that...

What we see is that we will be here for literally an infinite amount of time trying to write down the assignment for each element in ℕ, and this is because we have infinitely many assignments to make! Seeing as no one has an infinite amount of time to sit and write down infinitely many assignments, what we need to do is figure out a way to make infinitely many assignments in only a finite amount of time. Now, if the function $f : \mathbb{N} \to B$ is totally random — i.e., there is absolutely no pattern to which numbers go to a and which numbers go to b — then it will genuinely be very difficult to specify where each number is sent. However, suppose we want to define a function $f : \mathbb{N} \to B$ that has some nice pattern. For example, suppose we wanted to define the function that sends every even[2] number in ℕ to a in B, and every odd number to b in B. We note that this is a perfectly good function since every element in ℕ is either even or odd,[3] and thus every element in ℕ gets sent to one and only one element in B.

In order to write this function down, however, we need to get a bit creative. In particular, we cannot simply write $f(1) = b$, $f(2) = a$,

---

[1] In some books, the natural numbers include 0, so that the natural numbers correspond to the set $\{0, 1, 2, 3, ...\}$. We, however, will stick with our definition of the natural numbers as being the numbers $\{1, 2, 3, ...\}$, without zero. If we ever want to deal with the set $\{0, 1, 2, 3, 4, ...\}$, then we will denote this set by something like $\mathbb{N}_0$, where the "0" subscript reminds us that $\mathbb{N}_0$ includes the number zero. Again, though, $\mathbb{N} = \{1, 2, 3, 4, ...\}$ is the set of natural numbers for us in these volumes.

[2] Recall that every whole number is either even (like 2, 4, 6, 8, 10, and so on — i.e., any number that is divisible by 2) or it is odd (like 1, 3, 5, 7, 9, and so on — i.e., any number that is only one away from an even number).

[3] In particular, no element in ℕ is both even **and** odd, and no element in ℕ is **neither** even **nor** odd.

$f(3) = b$, $f(4) = a$, and so on, because we would be here all day (and all day tomorrow, and the next day, and the next day, forever). What we **can** do, however, is simply define our function a bit more abstractly, by saying that "$f : \mathbb{N} \to B$ is the function defined by $f(x) = a$ if x is an even number, and $f(x) = b$ if x is an odd number." We now immediately know where every element in $\mathbb{N}$ goes under the action of the function f. In particular, our definition tells us that if the enemy hands us some element x in the set $\mathbb{N}$ and wants us to tell him where it gets sent by f, then we simply need to find out if x is an even or odd number and respond accordingly. If we examine this element x and see that it is even, then we know it gets sent to a in B, and if it is odd then we know it gets sent to b. We therefore have been able to determine where any element in $\mathbb{N}$ goes without having to write down infinitely many assignments.

We can do even better though, and define the function f — which sends even numbers in $\mathbb{N}$ to a in B and odd numbers in $\mathbb{N}$ to b in B — in an even more mathematical and symbolic way. To do this, we need to first reconsider how we describe even and odd numbers. Suppose x is some number in $\mathbb{N}$, and suppose that x is even. We then know that x is divisible by 2, and so we can write[4] $x = 2m$ where m is some **other** number in $\mathbb{N}$. For example, if $x = 6$ then $m = 3$ because $x = 6 = 2m$ so that m equals 6 divided by 2. Similarly, if $x = 20$, then $m = 10$, and if $x = 36$ then $m = 18$. We therefore see that **a number x in $\mathbb{N}$ is even if and only if $x = 2m$ for some other number m in $\mathbb{N}$**.

We can say a similar thing about odd numbers. In particular, suppose x in $\mathbb{N}$ is an odd number. We then know that x must be one away from an even number. And since an even number can be written as $2m$ for some number m in $\mathbb{N}$, then we know that any odd number x in $\mathbb{N}$ can be written as $2m - 1$, since this is one away from the even number $2m$. For example, if $m = 1$ then $2m - 1 = 1$, if $m = 2$ then $2m - 1 = 3$, if $m = 3$ then $2m - 1 = 5$, and so on. We therefore see that as the number m goes through the numbers $1, 2, 3, 4$, and so on, the number $2m - 1$ goes through the numbers $1, 3, 5, 7$, and so

---

[4]Recall that when we write something like $x = 2m$, we mean "x equals the number obtained by **multiplying** 2 by m." I.e., when we simply stick two numbers next to each other, it means that we are multiplying them.

on. In this way, we can now say that **a number x in $\mathbb{N}$ is odd if and only if $x = 2m - 1$ for some other number m in $\mathbb{N}$.**

An important question that one might ask is why we did not choose to write an odd number x as $2m + 1$ (instead of $2m - 1$), since $2m + 1$ is **also** precisely one away from an even number. The answer to this question is that we indeed **can** write an odd number as $2m + 1$, but then we need to include the possibility that $m = 0$. This is because if m can only come from the set $\mathbb{N} = \{1, 2, 3, ...\}$, then $2m + 1$ can only take values $3, 5, 7$, and so on. These are indeed odd numbers, but we are missing the number 1, which is also an odd number. Thus, in order to write the number 1 as $2m + 1$, we need m to equal 0. However, zero is **not** in the set $\mathbb{N}$. Therefore, we **need** to use the minus sign instead of the plus sign in order to correctly say that "a number x in $\mathbb{N}$ is odd if and only if it can be written as $2m - 1$ for some other number m in $\mathbb{N}$." If we let $\mathbb{N}_0 = \{0, 1, 2, 3, ...\}$, then we can correctly say "a number x in $\mathbb{N}$ is odd if and only if it can be written as $2m + 1$ for some other number m in $\mathbb{N}_0$." For now, though, we stick with the first way of saying it, so that we only have to consider the set $\mathbb{N}$ and not the set $\mathbb{N}_0$.

We can now finally define the function $f : \mathbb{N} \to \{a, b\}$ which sends even numbers in $\mathbb{N}$ to a in $B = \{a, b\}$, and odd numbers in $\mathbb{N}$ to b in B. In particular, we say that given some x in $\mathbb{N}$, we have that $f(x) = a$ if $x = 2m$ for some other m in $\mathbb{N}$, and that $f(x) = b$ if $x = 2m - 1$ for some other m in $\mathbb{N}$.

What we have therefore seen in this section is that in order to explicitly define a function whose domain has infinitely elements, it is helpful to find a way to make infinitely many assignments with only a finite number of words or symbols. The statement "$f(x) = a$ if $x = 2m$ for some other m in $\mathbb{N}$" is one such way of doing so, and throughout these volumes we will see many more.

The Hilbert Hotel

Now that we have seen one way in which infinite sets are different from finite sets (namely, in how one can define functions from them), we now turn our attention to another strange phenomenon that arises with infinite sets. In order to see this strange phenomenon for infinite sets, let us first examine the **lack** of this strange phenomenon for finite sets. Let us consider the following scenario. Suppose there is a hotel called the Hilbert Hotel,[5] and suppose that there are 30 rooms in this hotel. In other words, suppose we have a set of 30 rooms, and let us label each room by the letter R, so that room 1 is labeled R_1, room 2 is labeled R_2, and so on. Suppose we called this set of rooms $ROOMS$, so that we have $ROOMS = \{R_1, R_2, R_3, ..., R_{29}, R_{30}\}$.

Now suppose we have a set of exactly 30 guests that want to stay in the Hilbert Hotel, and let us denote this set of guests by $GUESTS = \{G_1, G_2, G_3, ..., G_{29}, G_{30}\}$, where G_1 corresponds to guest 1, G_2 corresponds to guest 2, and so on. Being the good hotel managers that we are, we decide that we want to give each guest their own room. To do this, we need to come up with a function that assigns a different room to each guest. For simplicity, let us take the function that sends guest 1 to room 1, guest 2 to room 2, guest 3 to room 3, and so on, until finally we send guest 30 to room 30. Symbolically, our function f from $GUESTS$ to $ROOMS$ (so that $f : GUESTS \to ROOMS$) is defined by the assignments $f(G_1) = R_1$, $f(G_2) = R_2$, $f(G_3) = R_3$, and so on, finishing with $f(G_{30}) = R_{30}$. We note that this function f is a bijective (and therefore injective as well as surjective) function from the set $GUESTS$ to the set $ROOMS$, since we have been able to fill each and every room with precisely one guest. The following figure depicts how this function acts on the guests.

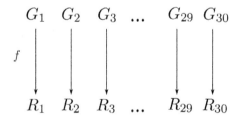

Figure 7.1: A depiction of how we fill up our 30-room hotel.

---

[5]We choose this as the name of our hotel because the mathematician who initially thought about such things was the great David Hilbert.

THE HILBERT HOTEL 71

Let us now consider the possibility that at the last minute, a friend of one of our guests decides to come to the hotel to visit, and that this friend also wants a room of his own.[6] Let us call this friend F_1. It is then clear that we cannot provide F_1 with a room of his own without kicking out one of our other guests, because our hotel is already fully booked. Since we are good hotel managers we most certainly do not want to kick out any of our current guests. In the mathematical language that we have set up so far, we must tell F_1 that he cannot stay at the hotel because we are unable to define an injective function from the set $\{F_1, G_1, G_2, G_3, ..., G_{30}\}$ to the set $\{R_1, R_2, R_3, ..., R_{30}\}$. This is because if we **could** define such an injective function, we would then have 31 people assigned to 30 rooms, without putting two people in the same room. This is obviously impossible.

Something remarkably different happens once we have a hotel with infinitely many rooms. Of course, no hotel can ever **actually** have infinitely many rooms, but that does not stop us from considering what would happen if a hotel **could** have infinitely many rooms.[7] Let us suppose now that our hotel has infinitely many rooms labeled R_1, R_2, R_3, R_4, and so on, forever. Thus, our set of rooms is now $ROOMS = \{R_1, R_2, R_3, R_4, ...\}$ where the "..." again means that the pattern continues on forever.

Let us also suppose that our hotel is completely filled with infinitely many guests, where we have labeled the guests G_1, G_2, G_3, and so on, forever. In other words, suppose that we now have a set of guests that we call $GUESTS = \{G_1, G_2, G_3, ...\}$. Being the good managers that we are, we have given each guest their own room, and we have assigned them as we did when our hotel had only 30 rooms. In particular, we assigned guest G_1 to room R_1, guest G_2 to room R_2, guest G_3 to room R_3, and so on. Writing this out in symbols (where f is again the function that sends people to the room to which they are assigned), we can make the **infinitely many** assignments $f(G_i) = R_i$, which simply means that we are sending the

---

[6] In other words, suppose this friend is very choosy and does not want to sleep on the couch of his friend's room.

[7] Indeed, one of the most remarkable features of mathematics is that it allows us to ponder ideas that have absolutely no physical relevance at all, just like a hotel with infinitely many rooms.

i^{th} guest[8] to the ith room.

We should reflect for a moment on the fact that according to this way of distributing our guests, every single room in our infinitely large hotel is occupied. Namely, we could go into **any** room at all and see whether or not it is occupied, and the answer in this case will always be yes. This is simply because any room we look into will have some number — suppose it is $1,405$ — and so when we go to look in this room, room $R_{1,405}$, we will find guest $G_{1,405}$ occupying it. This is depicted in Figure 7.2, which shows the (infinitely large) arrow diagram corresponding to the "room assignment function" that we just defined.

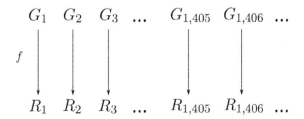

Figure 7.2: A depiction of how we fill up our infinitely large hotel. In particular, the "..." at the end corresponds to the fact that these assignments go on and on forever.

Now what happens if guest G_1 decides at the last minute to bring a friend, F_1, and that this friend wants his own room as well, separate from every other guest in the hotel? Recall that when our hotel was fully booked with only 30 rooms, we could not accommodate for this friend's wishes. And once again we find ourselves in a situation in which we have a fully booked hotel as well as a new individual wanting his own room.

Remarkably, we actually **can** now accommodate for this friend, without having to build any more rooms! To see this, we note that what we want to do is send each element in the set $\{F_1, G_1, G_2, G_3,$

---

[8]Note that this simply means the 1st guest when $i = 1$, the 2nd guest when $i = 2$, the 3rd guest when $i = 3$, the 4th guest with $i = 4$, and so on. Since i can take any value in $\{1, 2, 3, 4, ...\} = \mathbb{N}$, this corresponds to infinitely many assignments.

THE HILBERT HOTEL

G_4, ...} to a different element in the set $\{R_1, R_2, R_3, ...\}$ (note that the set of rooms has not changed since before the friend F_1 arrived). What we can do now is send F_1 to the room R_1, and then send G_1 to the room R_2, then G_2 to the room R_3, then G_3 to the room R_4, and so on. In other words, we send F_1 to the room R_1 and then we "shift" all of the other guests up a room, by sending the ith guest G_i to the $(i+1)^{th}$ room G_{i+1}. Figure 7.3 depicts how we make these assignments. We have therefore found a way for every guest (and the friend) to get his or her own room even though prior to the friend's arrival the hotel was fully booked, and we did not even need to put anyone in a closet or the basement (or build a new room)! This is incredible, and is something that is completely dependent on the fact that our hotel has infinitely many rooms.

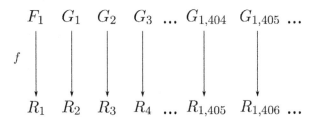

Figure 7.3: A depiction of how we can accommodate for the friend F_1 as well as the infinitely many guests $\{G_1, G_2, G_3, ...\}$.

Indeed, the world of the infinite gets even weirder! For instead of the guest G_1 only bringing his friend F_1, let us suppose that he actually brought 1,000 of his friends $F_1, F_2, F_3, ..., F_{1,000}$. Let us also suppose that each of these thousand friends are all equally spoiled, and so they all ask for their own room. On the surface of things it seems as though we have absolutely no hope of accommodating 1,000 new guests when our hotel was already fully booked, but indeed we can!

In order to do so — i.e., in order to accommodate for 1,000 more guests after our hotel was fully booked — all we need to do is send the first 1,000 friends (F_1 through $F_{1,000}$) to the first 1,000 rooms, and then send the infinitely many guests ($G_1, G_2, G_3, ...$) to the remaining rooms. In other words, instead of just "shifting" the guests up one room as we did when only one friend was added, we are now simply "shifting" the guests up by 1,000 rooms, and we can do this because we have infinitely rooms in the hotel! In particular, the function f which assigns people (i.e., the infinitely many guests as well as the friends of G_1) to their respective rooms could be given by $f(F_1) = R_1$, $f(F_2) = R_2$, $f(F_3) = R_3$, and so on, until we have $f(F_{1,000}) = R_{1,000}$. Then, once we are done giving these 1,000 friends of G_1 their rooms, we start with the guests G_1, G_2, G_3, and so on. To do so, we simply send guest G_1 to the next room, namely room $R_{1,001}$, and guest G_2 to room $R_{1,002}$, and so on. Thus we have $f(G_1) = R_{1,001}$, $f(G_2) = R_{1,002}$, $f(G_3) = R_{1,003}$, and so on, forever. In short, we are sending each guest to the room whose number is the same as the number of the guest, plus 1,000 (this is because the first 1,000 rooms are already taken by the friends). This is depicted in Figure 7.4.

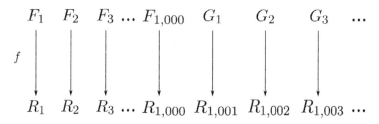

Figure 7.4: A depiction of how we can accommodate for one thousand friends $\{F_1, F_2, F_3, ..., F_{1,000}\}$ as well as the infinitely many guests $\{G_1, G_2, G_3, ...\}$.

Let us quickly recast the preceding paragraph in more symbolic notation, as it will help us in the next scenario when **even more** friends show up. Let us notice that we can define this function f — the function that sends the 1,000 friends to the first 1,000 rooms and the guests to the remaining rooms — as follows. We can write $f(F_i) = R_i$, where i is in the set $\{1, 2, 3, ..., 1,000\}$. All this is saying is that for **any** i that we choose in the set $\{1, 2, 3, ..., 1,000\}$, the function f sends F_i to R_i. This is of course what we want, since we know we want to send, for example, F_{307} (corresponding to $i = 307$) to the room R_{307}. Finally, to make the **infinitely many** assignments for the remaining infinite number of guests, we can write $f(G_i) = R_{1,000+i}$ where now i is in the set $\{1, 2, 3, 4, ...\} = \mathbb{N}$. All this is saying is that for any choice of i in the set $\mathbb{N}$ (i.e., for any of the infinitely many guests G_i), we are sending each guest to the room whose label is 1,000 more than the label of the guest. As an example, we would be sending G_{48} (corresponding to $i = 48$ in the set $\mathbb{N}$) to the room $R_{1,048}$ (since $1,000 + i = 1,048$ when $i = 48$). In this way we have seen how we can accommodate for a thousand friends of one guest even when the hotel was initially fully booked. This is indeed remarkable, but in fact we can still do better!

Suppose now that instead of the guest G_1 inviting only 1,000 friends, guest G_1 actually invites **infinitely many** of his friends. Let us label these friends as before, but now the list of friends never ends. We therefore have a set of (infinitely many) guests $\{G_1, G_2, G_3, ...\}$ and a set of (infinitely many) friends $\{F_1, F_2, F_3, ...\}$, and let us also suppose that each of these infinitely many friends wants their own room. Remarkably, we can indeed accommodate for all of this!

To do so, we simply need to send F_1 to room R_1, **and then skip room** R_2 and send F_2 to room R_3, and then skip R_4 and send F_3 to room R_5, and so on. In other words, we are simply sending **each** friend F_1, F_2, F_3, and so on, to the odd-numbered rooms R_1, R_3, R_5, and so on. In this way, we have not only been able to give each friend his own room, but we have also been able to keep the infinitely many even-numbered rooms completely available! Namely, after assigning the friends to their rooms, we notice that the rooms R_2, R_4, R_6, and so on, are still empty. Thus, we can send G_1 to the room R_2, and G_2 to the room R_4, and G_3 to the room R_6, and so on. In doing so, we will have given each guest their own room as well, and no two people (guest or friend) will have to share a room! Figure 7.5 depicts how these assignments work.

Let us now write all of this out more mathematically. Thankfully, our work in the previous section will make this easier on us. In particular, what we notice is that our "room assignment function" f sends the friends F_i to rooms according to the relation $f(F_i) = R_{2i-1}$. What this means is that each friend goes to the room whose label is two times the label of the friend, minus 1 (so that this room number will be an odd number). Thus, when $i = 1$, we see that $2i - 1 = 1$ (because 2 times 1 is 2, and then we subtract by 1). Similarly, when $i = 2$ we see that $2i - 1 = 3$, when $i = 3$ we see that $2i - 1 = 5$, and so

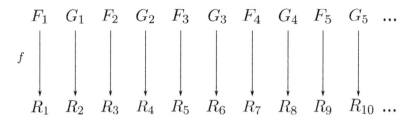

Figure 7.5: A depiction of how we can accommodate for infinitely many friends $\{F_1, F_2, F_3, ...\}$ as well as the infinitely many guests $\{G_1, G_2, G_3, ...\}$.

on, so that indeed F_1 goes to room R_1, F_2 goes to room R_3, F_3 goes to room R_5, and so on. Finally, we see that our "room assignment function" f sends the guests G_i to rooms according to the relation $f(G_i) = R_{2i}$. In particular, our function sends each guest to the room whose label is two times the label of the guest (and this room number will therefore be an even number). As examples, we see that guest G_1 gets sent to room R_2, since 2 is 2 times 1. Similarly, we see that guest G_2 gets sent to room R_4, that guest G_3 gets sent to room R_6, and so on. And there we have it — we have been able to send each guest **and each of the infinitely many friends** to their own rooms, despite the fact that the hotel was initially fully booked!

We saw initially that none of this "adding in friends" was possible when our hotel had only a finite number of rooms. Therefore, all of these ways of jamming in friends while allowing every person in the hotel to have their own room relies heavily on the fact that the hotel had infinitely many rooms. This is just one example of how weird the concept of infinity really is. Indeed, it still gets even weirder! We can, in fact, accommodate for the case when **each one of the infinitely many guests** invites infinitely many friends. We will see that we can do so while giving every single person their own room, and we can even do so while leaving open infinitely many rooms! We will explore how to do this in a later volume[9] when we have developed a bit more mathematical machinery. For now and for the next several chapters, we turn to developing more machinery for dealing with infinity.

---

[9]Though the reader is of course encouraged to think about how to do this with the machinery that we have developed so far.

Chapter 8

Bijectivity And Counting

A Return To Injectivity And Surjectivity

It might be surprising that the word "counting" does not show up until the eighth chapter of a book about mathematics. After all, what could mathematics be about if not counting? Recall, however, that we are **building** mathematics as we go. We do so by defining mathematical structures and proving certain statements about those structures. Nowhere in this process is it clear that counting things should be a fundamental process in what we are calling mathematics. We began by defining the abstract concepts of elements and sets, as well as the functions that relate different sets to each other. Before doing so, we had to let go of our preconceived notions of what mathematics is or is not about. Thus, whether or not the mathematics that we are developing has anything to do with counting, or "solving for x," or calculating some quantity, is simply not a worthwhile question.

It turns out, however, that all of the abstract ideas that we have been dealing with so far in this volume actually lend themselves very nicely for defining a notion of counting — in an abstract way — the number of elements in a set. Beginning with the next chapter, we will go on to use this new technique of counting to prove some truly remarkable things about infinity. In particular, we will find that infinity is infinitely more complex than we could have ever imagined!

For now, let us begin at the beginning. Recall that in Chapters 5 and 6 we introduced the concept of functions between sets. These were associations which "mapped" each element in one set to some element in another. To be precise, let us recall that a function f from a set A to a set B, written $f : A \to B$, sends each element in A to some element in B. Additionally, we recall that an injective function is a function that maps elements in its domain (here, the domain of f is A) to elements in its codomain (here, the codomain of f is B) in such a way that no two distinct elements in the domain get sent to the same element in the codomain. Recall that in our school dance analogy of Chapter 6, this occurs when there are never two boys dancing with the same girl. A surjective function is a function that "hits" every element in its codomain with at least one element in its domain. In our analogy, this occurred when every girl had at least one boy to dance with. A bijective function is simply a function that is both injective and surjective. We will explore this concept more now.

In particular, what does bijectivity have to do with counting? Let us first see what injectivity and surjectivity have to do with counting, since if we can understand these cases then we will certainly be able to understand the bijective case (which is just both at once). Let us continue on with the school dance analogy. We saw before that if we wanted to define a surjective function from the boys to the girls, we need **at least** as many boys as there are girls. This is because the definition of a function forces us to only assign one girl to each boy,[1] and not more. Thus, if there are more girls than boys, then even if we assign a different girl to each boy, we would still have girls left over. Therefore if there is some set of boys and some set of girls, and someone tells us that they have successfully defined a surjective function from the boys to the girls, then we automatically know that there are at least as many boys as there are girls. I.e., the number of boys is **greater than or equal** to the number of girls.

What if someone were to tell us that they successfully defined an injective function from the boys to the girls? What does this say about how many boys and girls there are? Well, we know that an

---

[1] This is assuming, as we did in Chapter 6 and as we are doing now, that the function is **from** the set of boys and **to** the set of girls. I.e., the boys are the domain of the function, and thus each boy must be sent to one and only one element in the codomain (the girls, in this case).

injective function sends each boy to a different girl. I.e., once one boy is assigned a girl, then no other boys get assigned to that girl. If it is possible to assign all the boys to girls in this way, then it must be the case that there are at least as many girls as there are boys. For suppose there are less girls than boys. As we start to assign boys to girls, we will "run out" of girls before we do boys, since we can never assign two boys to the same girl (by definition of injectivity). However, we must send every boy to **some** girl, by definition of a function.[2] Thus it must be the case that being able to successfully define an injective function from boys to girls is equivalent to having at least as many girls as there are boys. I.e., the number of boys is **less than or equal to** the number of girls.

We can now see what a bijective function between boys and girls means. Namely, if it is possible to define a bijective function from the boys to the girls, then there must be **just as many** boys as there are girls. For if the function is bijective, then it is surjective, and so the number of boys is greater than or equal to the number of girls. But we also know that if the function is bijective then it is also injective, and therefore the number of boys is less than or equal to the number of girls. Thus we have that the number of boys is greater than or equal to the number of girls, **and** that the number of boys is less than or equal to the number of girls. There is only one way for one number to be both "greater than or equal to" as well as "less than or equal to" some other number, and that is if the two numbers are the same. Thus, we know that if we are given a set of boys and a set of girls and someone tells us that they have managed to successfully define a bijective function from the boys to the girls, then we immediately know that there are exactly as many boys as there are girls!

We should also note that if there is some bijective function from the boys to the girls, then (assuming there is more than one boy and one girl) there are actually many **different** bijective functions from the boys to the girls. This is not surprising, for if there is one way to perfectly assign one boy to one girl without leaving anyone out, then there are in fact many ways to do so by simply "shuffling around" the various people involved. What is important is that there is **at least**

---

[2] As the previous footnote mentions, the boys are the domain of this function and therefore each boy must be sent to a girl.

one bijective function from the boys to the girls, for then (and only then) do we know that there are the same number of both genders.

The Most Inefficient Way To Count

Let us now return to what this all has to do with counting. Suppose we are handed a set of apples and that we are asked how many apples there are in the set. What would we do? Naturally, we would count them of course! But what are we **really** doing when we are counting them? In other words, how can we **rigorously** define the notion of "how many there are"? Now that we have the notion of bijectivity in our heads, we can rather easily define the very act of counting in a **completely abstract and rigorous** way.

Consider the set $\{1, 2, 3, 4, ..., N\}$ where N is some positive whole number, and where the "..." means that all the positive whole numbers between 4 and N are also included in the set.[3] For concreteness, consider the set $\{1, 2, 3, 4, 5\}$, i.e., when $N = 5$. Clearly, if we have a set of 5 apples, then we can define a bijective function from $\{1, 2, 3, 4, 5\}$ to our apples. Additionally, we most certainly could **not** define a bijective function from $\{1, 2, 3, 4\}$ to our set of apples, nor could we define one from $\{1, 2, 3, 4, 5, 6\}$ to our set of apples. It has to be $\{1, 2, 3, 4, 5\}$. This may seem obvious, especially because we knew ahead of time that our set of apples has 5 elements in it. After all, we said it was "a set of 5 apples." But let us suppose that we naively do not know how to count. After all, if we are truly building mathematics from the ground up, we **should not** know how to count, since we have not defined counting yet. We therefore have to **define** counting in some precise way. In other words, we need to give **precise** meaning to the phrase "a set A has N elements in it." Let us therefore make the following definition.

Definition 8.1. Given some set A, we say that A has N elements in it if there is a bijective function from the set $\{1, 2, 3, ..., N\}$ to the set A. We then say that the set A has **cardinality** N.

---

[3]For example, if $N = 11$, then the set $\{1, 2, 3, 4, ..., N\}$ would be $\{1, 2, 3, 4, 5, 6, 7, 8, 9, 10, 11\}$. We can quickly see that the notation $\{1, 2, 3, ..., N\}$ is much more convenient than explicitly writing out the elements, especially if $N = 1,562,351$, for example.

Let us look at a couple of quick examples. We see that the set $A = \{a, b\}$ has cardinality 2 since we can define the bijective function $f : \{1, 2\} \to A$ by $f(1) = a$, $f(2) = b$. We also note that the function $g : \{1, 2\} \to A$ defined by the assignments $g(1) = b$, and $g(2) = a$ is also a bijective function. Thus, it does not matter that there is **more than one** such bijective function — all that matters is that there is **at least** one. As a final example, we see that the set $A =$ {Kobe, LeBron, Star Wars, 8} has cardinality 4 since we can define the bijective function f from the set $\{1, 2, 3, 4\}$ to A by the assignments $f(1) =$ Star Wars, $f(2) =$ Kobe, $f(3) =$ LeBron, and $f(4) = 8$. We finally note the obvious fact that if some set has cardinality N, then it cannot **also** have any other cardinality. Namely, if there is a bijective function from the set $\{1, 2, 3, 4, 5\}$ to some set A, then there is **not** a bijective function from the set $\{1, 2, 3, 4, 5, 6\}$ to the same set A. This is a very good thing, for it reflects the fact that the same basket of apples cannot simultaneously have exactly 5 and exactly 6 apples in it! If this were **not** the case and our definition of counting allowed for the possibility that one set could simultaneously have two different cardinalities, then we would want to change our definition of counting!

We could have also chosen some other set with N elements in it when defining what it means for a set to have N elements in it. In other words, we could say "a set A has 3 elements if there is a bijective function from the set $\{a, b, c\}$ to A," or we could say "a set A has 4 elements when there is a bijective function from {apple, goose, Mars, grape drink} to A." It is simply a convention (and a good one) to use the set $\{1, 2, 3, ..., N\}$ for some whole number N. Thus, if we want to decide how many elements are in some set, we simply start with the set $\{1\}$ and ask if there can be a bijective function from $\{1\}$ to our set. If there cannot be such a function, then we move on to $\{1, 2\}$ and ask the same question. If the answer is no, we move on to $\{1, 2, 3\}$, and then $\{1, 2, 3, 4\}$, and so on, until we finally find that there can be such a function. We would then know that no other set of this form[4] will do, so we simply look at which "N" we stopped at and conclude that that is how many elements are

---

[4]I.e., of the form $\{1, 2, 3, ..., N\}$ for some N.

in our set!

Of course, this is not how we count in real life, but this is because we usually do not care much about the world of abstraction and rigor when we are buying apples at the grocery story. However, the fact that we can use this method to define "counting" rigorously has a significance that cannot be overstated. Its power and beauty will become evident in the next couple of chapters as we begin to see how perfect of a tool this method is for analyzing all the different types of infinity. And yes, there is more than one type of infinity!

Chapter 9

Counting Infinity

A Problem With Our Counting

In the previous chapter we saw how we could use our notion of bijectivity to give a rigorous meaning to our intuitive idea of counting the elements in a set. Namely, if we have a set A and there was some N (a positive whole number) such that the set $\{1, 2, 3, ..., N\}$ could be mapped bijectively to our set A, then we define our set to have N elements. It is important to note the flow of logic in this construction. We make no reference to "how many" elements are in A before we start looking for bijections. Instead, we simply consider the sets $\{1\}$, $\{1, 2\}$, $\{1, 2, 3\}$, $\{1, 2, 3, 4\}$, and so on, until we find one that lets us define a bijective function from it to our set. Recall that if we find an N that works, then there will be several different bijections from $\{1, 2, 3, ..., N\}$ to our set. But we also recall that this is perfectly fine. All that matters is that there is at least one! If we can find one such set — i.e., if we can find an N that works — then no other set of that form will do. In particular, if we find such an N (such that we can define a bijective function from $\{1, 2, ..., N\}$ to our set), then we know that this is the only N that will work.[1]

This all seems like a huge amount of work and abstraction in order to do something as trivial as counting. However, we will see in this

---

[1] It is for this reason that we know that the statement "set A has N elements" is well-defined, and that it cannot be the case that, for example, a set can simultaneously have 6 and 7 elements.

chapter that this way of doing things is actually immensely powerful, and it will allow us to make conclusions about our notion of infinity that we never could have made without these tools at our disposal.

To begin to uncover some of these remarkable truths about infinity, we must reflect upon when the machinery that we developed in the previous chapter does **not** work. In particular, one very important subtlety about our method of counting is that it simply does not work if the set A that we were given (and that we are trying to count) has infinitely many elements. This is because if we are given a set with infinitely many elements, then by the very nature of infinity, there will be no finite number N such that we can define a bijective function from the set $\{1, 2, 3, ..., N\}$ to the set A. For if A has infinitely many elements, then $\{1\}$ has no bijective function from it to A, nor does $\{1, 2\}$, nor does $\{1, 2, 3\}$, nor does $\{1, 2, 3, ..., N\}$ for any N!

Let us take the main example of an infinite set that we have dealt with so far and let $A = \mathbb{N}$, where $\mathbb{N}$ is the set of natural numbers that we introduced in Chapter 7. In particular, we are letting $A = \{1, 2, 3, 4, ...\}$ where we recall that the "..." means that this set "goes on forever," including only the positive whole numbers. Clearly, there is no N such that[2] the set $\{1, 2, 3, ..., N\}$ can be put into bijection with A, simply because $\{1, 2, 3, ..., N\}$ "stops" somewhere (at N), whereas A does not.

That is a bummer. Is it the case then that there are sets that we simply cannot count? In other words, are there sets that are immune to the high-powered weapon of counting that we just developed? If the answer to this question were yes then we probably would not have the current chapter in this volume at all, so we can guess that there is a way to adapt our method of counting even to infinite sets. What this problem is telling us is that we simply need to make our weapons more powerful.

And how do we do that? What we need to do is extend our method of counting to allow for these problematic cases of infinite sets. In particular, we need to extend our very notion of what a

---

[2] Let us not confuse the two symbols N and $\mathbb{N}$. They look different for a reason — the former is simply some finite number, whereas the latter denotes the entire set of natural numbers.

FIXING THE PROBLEM 85

number is in order to "count" sets with infinitely many elements. Our method of attack will be to do so in the most obvious way possible. As a disclaimer, it may be the case that what follows seems like a "stop-gap" of sorts. However, the results that we will derive in this and the following couple of chapters will show that there is in fact some kind of deep truth to all that we are developing.

Fixing The Problem

In order to extend our method of counting, we need to extend our very notion of what a number is. And how might we extend our numbers? Easy! As before, we say that if there is a bijective function from the set $\{1, 2, 3, ..., N\}$ for some number N to the set A, then the set A has N elements. Now all we need to do is say that if there is a bijective function from the set $\{1, 2, 3, ...\}$ (where now we note that this set goes on forever) to the set A then the set A has "infinity type 1" elements. In other words, we are literally defining a **new number** and calling it "infinity type 1." Some numbers are finite (in fact, a lot of numbers are finite), and some numbers are infinite. So far, we have only seen one infinite number — we have just introduced it and called it "infinity type 1." In order for us to know whether or not a given set A has "infinity type 1" elements in it, we simply need to see whether or not it is possible to define a bijective function from the set $\{1, 2, 3, ...\}$ to the set A. It is important to note how similar this process is to counting[3] a set with finitely many elements.

In order to avoid extremely awkward language, let us make this a formal definition. Note that we are **defining** what we mean by "infinity type 1" here, and that there is nothing to stop us from doing so. We include the definition of cardinality from the previous chapter simply to emphasize the fact that we are simply **extending** our notion of what it means for some set to have some number of elements in it, where now "some number" might be the number "infinity type 1."

Definition 9.1. If for some N there is a bijective function from the

---

[3]Where here, by "counting" we mean using the new and powerful machinery that we developed in the previous chapter.

set $\{1, 2, ..., N\}$ to the set A, then we say that A has cardinality N. If there is a bijective function from the set $\mathbb{N} = \{1, 2, 3, ...\}$ to the set A, then we say that A has cardinality "infinity type 1."

Thus, the cardinality of a set is just the number of elements in that set, where "infinity type 1" is viewed as some new infinite number. Recall from the Note On Notation at the beginning of this volume that "3" is just a symbol for the abstract idea of "the number three." In our case this translates to the abstract idea of "a set's ability to have a bijective function be defined from the set $\{1, 2, 3\}$ to it." In the same exact way, we view "infinity type 1" simply as a symbol (just like "3" or "16" or "1,245,695") and what that symbol stands for is the abstract idea of a set's ability to have a bijective function be defined from $\{1, 2, 3, ...\}$ to it.

One may now be wondering why we are calling this new number "infinity type 1," as opposed to just "infinity." Is there an "infinity type 2"? What about type 3? Surely there cannot be more than one type of infinity, right?

Let me take this question as an opportunity to reflect upon my own personal upbringing. I recall being on the playground as a younger child (in the first grade) and arguing with a buddy about who a girl liked more — me or him. I argued that she liked me more, and he argued that she liked him more times 2. After some quick analysis of the data, I concluded that she in fact liked me more times a million, but was immediately dealt the death blow when my buddy informed me that she in fact liked him more "times infinity." That was it, I was toast. My first grade crush was forever lost, simply because there was nothing I could say that could bring my "likeability" up over infinity. The simple reason for this is that infinity plus anything is still infinity, and infinity times anything is still infinity.

What a poor, unfortunate first grader I was, for I simply did not know at the time that I was in fact **not** out of the fight for my beloved! While it is of course true that anything times infinity is still infinity, it is **not** true that there is nothing greater than infinity. In fact, what I should have asked my competitor was **which** infinity he was talking about, because no matter which infinity he was talking about, I would be able to find one that is greater! Unfortunately, I still would not have necessarily won[4] because this back-and-forth of greater and greater infinities in fact can go on forever! This is because there is an infinitely high tower of, reaching up into the heavens, of ever-increasing infinities!

This might all sound more like fantasy than mathematics, but throughout the next six chapters we will make all of this completely clear and rigorous, thus equipping us for our next verbal battle for love. In the next chapter, though, we will take on the much more modest task of showing that infinity plus anything is indeed "the same" infinity, and that "infinity times 2" is also the same infinity.[5] We will then go on to show that even "infinity times infinity" is still "the same" infinity. We will continue to press on though, against all odds, and actually uncover this infinite tower of new and ever-increasing infinities!

---

[4] I would, however, have at least been able to extend the fight for my beloved.
[5] In doing so, we will also make precise what we mean by "the same infinity."

Chapter 10

Infinity Times 2

Being Precise About Infinity

Last chapter we used our abstract way of counting the number of elements in a set to extend the "normal" number system to include a number which we called "infinity type 1." In particular, analogously to how we define a set A to have N elements if we can define a bijective function from the set $\{1, 2, 3, ..., N\}$ to A, we also define a set A to have "infinity type 1" elements if we can define a bijective function from the set $\{1, 2, 3, ...\}$ to A. Implicit in our calling this new number "infinity type 1" is that there is likely an "infinity type 2." For if there were not an "infinity type 2" we may as well have called "infinity type 1" simply "infinity." And indeed, there is more than one type of infinity (many more types, in fact), but before we can truly appreciate this fact we need to first get used to what this definition of "infinity type 1" really means.

We all already know that "infinity plus anything is infinity" and "infinity times anything (other than 0) is infinity," and other sorts of "obvious" statements like these. However, if we claim that there are different kinds of infinity (which is what we are claiming), then we need to be more careful when we make statements like "infinity plus anything is infinity." In particular, we need to know **which kind** of infinity plus anything is **which kind** of infinity.

In other words, suppose for a second that there is an "infinity type 2" (there is, but since we have not proved it yet let us just suppose it exists). Then we need to know whether "infinity type 1" plus anything is "infinity type 1" or if "infinity type 1" plus anything is "infinity type 2." Or maybe it is the case that there are some things that we could add to "infinity type 1" to give us "infinity type 1" again, but that there are other things that we could add to "infinity type 1" which would give us "infinity type 2."

While all of this might sound a bit strange or a bit contrived, we will soon see that these ideas are in fact completely rigorous. Indeed, the rigor and beauty of these ideas are such that we will be forced to somehow believe that there "really are" many kinds of infinities.[1] For now this is enough of the fluffy philosophy — let us do some math. What we will focus on in this chapter is giving precise meaning to the phrase "infinity times 2 is infinity." What we will show is that the precise (and true) statement is that "infinity type 1" times 2 is "infinity type 1."

Infinity Times 2

Let us define a set that has, in a very obvious way, twice as many elements as a set with "infinity type 1" elements in it.[2] Then such a set would have "'infinity type 1' times 2" elements in it. Namely, let us define a set by taking $\{1, 2, 3, ...\}$, which has "infinity type 1" elements in it, and "creating another copy" of every element in it. We can make this rigorous by adding into our set every single negative number. Thus if $A = \{1, 2, 3, ...\}$, then let us define $B = \{1, -1, 2, -2, 3, -3, 4, -4, ...\}$ and so on. The set B is in some clear sense "twice as big" as A, since we define B by essentially adding another copy of A to itself. Namely, for every element in A, there is both a positive and a negative version of that element in the set B.

Recall, however, that our idea of what "twice as big" means now has a very precise meaning that we must respect. Namely, suppose

---

[1] Scare quotes are employed around "really are" because we will not be discussing the philosophical — almost spiritual — issues that arise when asking about the sense in which mathematical structures "really exist."

[2] Recall that we are viewing "infinity type 1" as a perfectly good (infinite) number, so that the phrase "set A has 'infinity type 1' elements in it" is indeed grammatically correct.

we have a set called $SET1$ that has cardinality N, so that there is a bijective function from $\{1, 2, 3, ..., N\}$ to $SET1$. Suppose also that we have another set called $SET2$ that has cardinality M, so that there is a bijective function from the set $\{1, 2, 3, ..., M\}$ to $SET2$. Then $SET1$ is "twice as big" as $SET2$ if and only if $N = 2M$.

Having this rigorous notion of what "twice as big" means might seem a little too abstract when comparing sets with only finitely many elements, but as we saw with the hotels in Chapter 7 our intuition is called into question once we have sets with infinitely many elements. Luckily, now that we have such a powerful and abstract method of counting at our disposal, we do not have to worry about our **intuitive** notion of what "twice as big" means, and instead we can just inquire about our rigorous definition of cardinality. Namely, let us ask whether or not $A = \{1, 2, 3, ...\}$ and $B = \{1, -1, 2, -2, 3, -3, ...\}$ have the same cardinality. The claim is that they **do** in fact have the same cardinality, but before proving this let us reflect on why this is a non-trivial statement.

One part of us wants to say that A and B have the same cardinality because two times infinity is still infinity. This was the reasoning that had me defeated in the battle for my beloved on the playground in first grade. However, the other part of us wants to say they do **not** have the same cardinality because B clearly has "more" elements than A. Let us therefore put an end to this debate.

In order to prove that A and B do have the same cardinality, we need to find a bijective function from the set $\{1, 2, 3, ...\}$ to the set $B = \{1, -1, 2, -2, 3, -3, ...\}$. This will prove that B has "infinity type 1" elements in it, and since we know that A also has "infinity type 1" elements in it (since $A = \{1, 2, 3, ...\}$ is the exact same set (namely, the set $\mathbb{N}$) that we used to **define** what it meant for a set to have "infinity type 1" elements in it) then we will have shown that A and B indeed do have the same cardinality. As we will see, defining a bijective function from $\mathbb{N} = \{1, 2, 3, ...\}$ to B is actually quite simple — it is virtually identical to how we managed to fit infinitely many friends and infinitely many guests into our hotel in Chapter 7!

First, let us take the subset of $\mathbb{N}$ that contains only the odd positive whole numbers — namely, let us form the subset $\{1, 3, 5, 7, ...\}$. Let us also do the same for the even numbers — namely, let us form the subset of ℕ consisting of $\{2, 4, 6, 8, ...\}$. All we need to do is define our function to send the odd elements in ℕ to the positive elements in B, and the even elements in ℕ to the negative elements in B. In other words, if F is our function from ℕ to B, then let us define the assignments $F(1) = 1$, $F(3) = 2$, $F(5) = 3$, $F(7) = 4$, and so on, sending the next highest **odd number** in ℕ to the next highest **positive number** in B. Similarly, let us define the assignments $F(2) = -1$, $F(4) = -2$, $F(6) = -3$, $F(8) = -4$, and so on, sending the next highest **even number** in ℕ to the next **lowest negative** number[3] in B. The function $F : ℕ \to B$ that we defined in this paragraph is depicted using arrows in Figure 10.1.

Let us note how similar this function is to the function that we defined in Chapter 7 when we had to fit infinitely many friends and infinitely many guests into our hotel. With a simple relabeling of the elements, this is completely identical to that discussion. In particular, if we look at Figure 10.1, we see that if we reverse the direction of the arrows, identify the positive numbers i in B with the friends F_i in Figure 7.5, identify the negative numbers $-i$ in B with the guests G_i (also in Figure 7.5), and finally identify the numbers in ℕ with the room numbers, then we have recovered the **exact same** function that we defined in our hotel discussion!

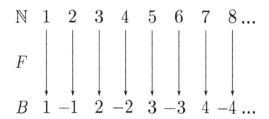

Figure 10.1: A depiction of the (bijective) function $F : ℕ \to B$ which shows that B and $A = ℕ$ have the same cardinality.

We now need to show that this function F is bijective, but this is

---

[3] We could have instead sent the odd numbers in ℕ to the negative elements in B and the even numbers in ℕ to the positive numbers in B. This would simply give yet another bijective function from ℕ to B, but for absolutely no reason at all we decided to make the choice that we did above.

actually quite clear. We see that no two elements in ℕ get mapped to the same element in B, and so F is injective. Additionally, since every element in B is either positive or negative, every element in B is eventually "hit" by something in ℕ. Therefore F is surjective. Thus, F is bijective, and therefore the set B has "infinity type 1" elements. Since our set A also has "infinity type 1" elements, we see that A and B have "the same number" of elements — namely, they have the "same infinity" of elements. We have thus proven that "infinity type 1" times 2 is "infinity type 1"!

Infinity Type 1 Plus Anything

We can use the same logic as above to quickly prove that[4] "infinity type 1" plus anything is "infinity type 1," where by "anything" we mean "any finite number." We can see this as follows. Let $A = \{1, 2, 3, ...\}$, as usual, and let $APPLES$ be a set of apples which contains some finite amount of apples. Since there are finitely many apples in $APPLES$, we know that there is some N such that there is a bijective function from $\{1, 2, 3, ..., N\}$ to the set $APPLES$. Let us therefore just label the apples in $APPLES$ as A_1, A_2, A_3, ..., A_N, so that $APPLES = \{A_1, A_2, A_3, ..., A_N\}$. Now we can formalize the notion of "'infinity type 1' plus anything" by considering the set with all of the elements in $A = \{1, 2, 3, ...\}$ **as well as** all of the elements in $APPLES$. This set represents a set with "infinity type 1 plus N" elements in it, since A has "infinity type 1" elements and $APPLES$ has N elements.

Let us denote this set of numbers and apples by

$$B = \{A_1, A_2, A_3, ..., A_N, 1, 2, 3, ...\}.$$

Now let us define a bijective function from ℕ to B, thus showing that B actually has "the same number of elements" as the set A itself — namely, "infinity type 1" elements. Again, we can let our discussion of infinitely large hotels in Chapter 7 guide us (where now instead of having to fit some finite number of friends into a previously full hotel,

---

[4] Note how we are already being sure to be precise about which infinity we are talking about.

we need to account for finitely many apples in a previously infinite set). In particular, let F now be the function from $\mathbb{N}$ to B defined by sending the first N numbers in $\mathbb{N}$ to the N apples in B. We then simply "shift" the rest of the numbers in $\mathbb{N}$ so that "$N+1$" in $\mathbb{N}$ gets sent to 1 in B, and "$N+2$" in A gets sent to 2 in B. In symbols, this would read

$$F(1) = A_1, \quad F(2) = A_2, \quad F(3) = A_3, \quad \ldots, \quad F(N) = A_N,$$

and

$$F(N+1) = 1, \quad F(N+2) = 2, \quad F(N+3) = 3,$$

and now we just continue that on forever! Figure 10.2 depicts this function, and again we see a similarity with our discussion of the infinitely large hotel — in this case the similarity is with our discussion of fitting 1,000 new friends into the hotel. In particular, by looking at Figure 10.2 we see that if we reverse the direction of the arrows and identify the N apples with the N friends,[5] the positive numbers i in B with the infinitely many guests G_i, and the numbers in $\mathbb{N}$ with the room numbers, then we get precisely the same function as that depicted in Figure 7.4. The function F that we defined above is clearly bijective, and so we have shown that "infinity type 1" plus any finite number is "infinity type 1"!

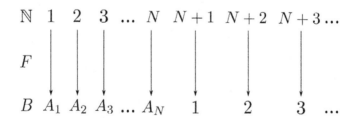

Figure 10.2: A depiction of the (bijective) function $F: \mathbb{N} \to B$ which shows that B and $A = \mathbb{N}$ have the same cardinality.

---

[5] In our hotel example, we arbitrarily chose to let $N = 1,000$, i.e. we arbitrarily decided to add in 1,000 friends. Our discussion of apples in this section shows that we could have chosen any number of friends!

We leave one more cool result for the exercises, providing lots of hints (as well as full solutions in the back of this volume). We need to get used to the ideas behind these types of functions[6] because we will need to get fancier and more clever with them in the next few chapters in order to prove some even cooler things. In the next chapter we will show that, in a very precise sense, "infinity type 1" times "infinity type 1" is "infinity type 1." The proof of this statement is absolutely gorgeous, but not nearly as gorgeous as the proof that we will see in Chapter 14 that there is in fact an infinity that is truly larger — i.e., more infinite — than infinity type 1.

So far all we have done is provide more evidence that there is only one infinity, since every infinite set that we have considered so far has had precisely "infinity type 1" elements. However, seeing the techniques used here and gaining respect for the logic that is at play is what we need to do before proving the existence of a whole infinitely large ladder of new infinities!

Exercises

1) Prove that "infinity type 1" divided by two is still "infinity type 1." Do this by realizing that the set of even numbers is a meaningful way of describing "infinity type 1 divided by two," since we can evenly split $\mathbb{N} = \{1, 2, 3, ...\}$ into odd and even parts. Thus, in a clear sense, the even numbers alone (as well as the odd numbers alone) have "half as many" elements as $\mathbb{N}$. Then define a bijective function from $\mathbb{N} = \{1, 2, 3, ...\}$ to $B = \{2, 4, 6, 8, ...\}$, thus showing that $\mathbb{N}$ and B actually do have the same cardinality!

---

[6]Namely, functions from $\mathbb{N}$ to some other infinite set in order to see if the infinite set has "infinity type 1" elements.

Chapter 11

Infinity Times Infinity

Setting The Stage

In the previous chapter we used our new weapons for counting and applied them to "counting" infinite sets. We saw that we can — in some sense — double the number of elements in a set with "infinity type 1" elements, and get a new set which also has "infinity type 1" elements. We also saw that we could add any finite number of elements to a set with "infinity type 1" elements, and still have a set with "infinity type 1" elements. Recall that whether or not a set still has "infinity type 1" elements is determined solely by whether or not we can define a bijective function from $\{1, 2, 3, ...\}$ to that set.

However, last chapter was merely a warm-up for the next few chapters. In particular, the statements that we will make about infinity in the next chapter and in Chapter 14, and the proofs of these statements, will take some more creativity and perhaps a bit more thought. In other words, the next few chapters will be harder than most. That being said, there are two things can be promised. The first is that it will not be all **that** bad, and the second is that the beauty of the results that we will uncover will make the process entirely worth it (as is usually the case with math).

The first thing we are going to do is show that the set of all fractions also only has "infinity type 1" elements in it. This is a truly remarkable result. However, in order to fully appreciate **why** this fact is so remarkable, we need to first remind ourselves about fractions and the nature of their infinity. Therefore this chapter will be spent laying the groundwork and refreshing our memory of fractions, and we will save the actual proof of the statement that there are only "infinity type 1" fractions for the next chapter.

A Quick Refresher

Recall that fractions are nothing but things like $\frac{1}{2}$ ("one-half", or "1 divided by two"), or $\frac{3}{4}$, or $\frac{2}{5}$. Namely, "$\frac{1}{2}$" is a symbol which represents the abstract idea of the number that, when multiplied by 2, gives 1. In the same way, "$\frac{3}{4}$" is a symbol which represents the abstract idea of the number that, when multiplied by 4, gives 3. Accordingly, we can think of fractions as nothing but a pair of whole numbers — the number that is "on top" and the number that is "on the bottom." Thus, in the fraction $\frac{3}{4}$, 3 is the number "on top" and 4 is the number "on the bottom." Given a fraction, if we multiply it by the number that is "on the bottom," then we get the number that is "on top." For example,

$$4 \times \frac{3}{4} = 3, \quad 2 \times \frac{1}{2} = 1, \quad \text{and} \quad 111 \times \frac{112}{111} = 112.$$

This might all be familiar, but if not then hopefully this suffices as a reminder. Let us now recall that we usually call the number "on top" the numerator, and the number "on the bottom" the denominator. Fractions are not scary at all when we recall that we deal with fractions all of the time. For example, a quarter of a basketball game is just $\frac{1}{4}$ of a whole game, a nickel is just $\frac{1}{20}$ of a whole dollar, and a cup might be $\frac{3}{8}$ full at some given time. The only difference between the fractions $\frac{1}{4}$ and $\frac{17}{1,045,231}$ is that we do not usually care when we are $\frac{17}{1,045,231}$ of the way through a basketball game, whereas we usually do care when we are $\frac{1}{4}$, $\frac{1}{2}$, or $\frac{3}{4}$ of the way through.

Another equivalent way to think about fractions — which is more in line with how we learned about fractions in school — is as follows. The fraction $\frac{3}{4}$ represents the number of pizzas everyone would eat if there were three pizzas and four people to distribute all of the pizza to (assuming we give everyone the same amount of pizza). Thus, if we have 17 pizzas and $1,045,231$ people want pizza, then every person gets "$\frac{17}{1,045,231}{}^{th}$" of a pizza. This is how we can see that zero is indeed a fraction, because "zero divided by anything is zero." Namely, if we have zero pizzas, then no matter how many people want pizza, everyone will get zero pizzas (because there simply is not any pizza to distribute). This is also one way that we can see why we **cannot** have a fraction with zero as the denominator. In other words, $\frac{17}{0}$ and $\frac{5}{0}$ are not fractions (nor is anything else with 0 in the denominator) because if we have some pizzas around and literally **zero** people to distribute them to, then the pizzas are kind of just "floating around" out there and nothing makes sense anymore. Thus, we simply cannot distribute 5 pizzas to 0 people, and so $\frac{5}{0}$ is not a well-defined fraction.

Thinking about pizza also helps us see that fractions can indeed be larger than 1. For instance, if we have three pizzas and only two people want to eat pizza, then both people get $\frac{3}{2}$ — or one and a half — pizzas. We can write one and a half as $\frac{3}{2}$ (as we just did), or we can also write it as $1 + \frac{1}{2}$, where this second way of writing one and a half is literally "one plus one half."

The last thing we need to recall here is that there is a notion of equality between certain fractions, even when these fractions might at first sight look different. Namely, we have that $\frac{1}{2}$ and $\frac{2}{4}$ are the same fraction — i.e., they are numerically equal. We also have that $\frac{1}{3}$ and $\frac{3}{9}$ are the same fraction, as are $\frac{2}{5}$ and $\frac{4}{10}$. The general idea is that if we have one fraction and we obtain another fraction by multiplying "the top" and "the bottom" by the same number, then the resulting fraction is actually the same as the one we started with. For example, $\frac{1}{2}=\frac{2}{4}$ because $\frac{2}{4}$ is obtained from $\frac{1}{2}$ by multiplying both the numerator and the denominator by 2. Similarly, $\frac{3}{9}=\frac{1}{3}$ because $\frac{3}{9}$ is obtained from $\frac{1}{3}$ by multiplying both the numerator and the denominator by 3. This makes sense, because if we have one pizza and three people to distribute them to, then everyone gets $\frac{1}{3}$ pizzas, and similarly if we have three pizzas and nine people to distribute them to, then everyone again gets $\frac{1}{3}$ pizzas.

We are now in a position to meaningfully define the set of **all fractions**. This set will include positive as well as negative fractions,[1]

---

[1] Just as in the case of whole numbers, a **negative** fraction is just a fraction but with a "−" sign in front of it. For now, a negative fraction can be thought of as "another copy" of its positive version. As we will see in a later volume, this can

and it also includes zero since zero is indeed a fraction (as we saw above). The set of all fractions is an infinite set, and as we will see in the next section it is a seemingly **very** infinite set. Let us for now set up some notation and agree to call the set of all fractions $FRAC$, so that we have $FRAC = \{$all fractions$\}$. Based on our discussion above, when we make this definition we are already implying that $\frac{1}{2}$ and $\frac{2}{4}$ are the **same element** in $FRAC$.

A Seemingly New Infinity

Now that our crash course on fractions is over, let us see what makes the fractions seemingly "more infinite" than the set $\mathbb{N} = \{1, 2, 3, ...\}$. We begin by noting that there is a **very important** fact about fractions that has no analog in the set $\{1, 2, 3, ...\}$. Namely, between any two **distinct** fractions, there are infinitely many more fractions! Seeing why this is **not** the case in $\{1, 2, 3, ...\}$ is easy. For suppose we were given two distinct elements — i.e., two different numbers — from the set $\{1, 2, 3, ...\}$. Let us call these two distinct numbers M and N for the sake of concreteness. Since M and N are not equal (by assumption), we know that one of these two numbers will be bigger than the other. And since we do not know what M and N are explicitly (other than that they are not equal), we can suppose that N is bigger than M. There are then clearly only finitely many elements between M and N. This is because all we need to do is count off M, $M+1$, $M+2$, until we hit N. Since both M and N are finite numbers, we will only have to count off finitely many elements.[2]

---

be made more precise, but for now it suffices to think of the negative sign as a somewhat formal object. In short, though, a negative fraction is simply a fraction that, when added to its positive counterpart, gives zero. Thus, $-\frac{1}{2} + \frac{1}{2} = 0$ in just the same way that, for example, $-5 + 5 = 0$.

[2] Let us look at an example. Suppose we chose the numbers $M = 17$ and $N = 1,000$. Then there are only 983 numbers between M and N, because $1,000 - 17 = 983$. This might all seem trivial right now, but a lot of interesting things in math come from seemingly trivial observations. Here, the **importance** of this trivial observation is that we can **count** the elements between M and N in the set $\mathbb{N} = \{1, 2, 3, ...\}$, and moreover the result of this counting will always be finite. As we will see shortly, we **do not** get a finite number (ever!) when we try to count the number of fractions between any two fractions.

A SEEMINGLY NEW INFINITY

Let us take a classic example now to see why there are infinitely many elements between any two distinct fractions. We begin by asking the following question. How many fractions[3] are there between $\frac{1}{2}$ and 1? The answer is that there are a lot of fractions between $\frac{1}{2}$ and 1, and right now we will only find infinitely many of them. In other words, we will find infinitely many fractions between $\frac{1}{2}$ and 1, and we will not have even found all of them!

The main idea is that there is a fraction "halfway" between $\frac{1}{2}$ and 1, as well as a fraction halfway between that fraction and 1, as well as a fraction halfway between that fraction and 1, and so on. We can do this infinitely many times without ever reaching 1, and so in this way we have found infinitely many fractions between $\frac{1}{2}$ and 1. To be precise, the fraction $\frac{3}{4}$ is halfway between $\frac{1}{2}$ and 1, and then the fraction $\frac{7}{8}$ is halfway between $\frac{3}{4}$ and 1, and then the fraction $\frac{15}{16}$ is halfway between $\frac{7}{8}$ and 1, and then the fraction $\frac{31}{32}$ is halfway between $\frac{15}{16}$ and 1. We can keep doing this forever without ever reaching 1, and thus we have uncovered infinitely many elements[4] between $\frac{1}{2}$ and 1 in *FRAC*.

The important point is that we can make this argument between **any two** fractions. In particular, if we are given any two distinct fractions, then all we need to do is step halfway from one to the other, then halfway again, then halfway again, and so on. Since each of these "halfway points" correspond to distinct fractions, we have uncovered infinitely many fractions between our two initial fractions.

Now here is the rub. Suppose the enemy hands us two distinct fractions. From the above considerations, we know that we can find infinitely many fractions between these two fractions. But in fact, we can do even better! For between any two of the infinitely many fractions that we have now found, there is yet **another** infinity of fractions! For example, one of the fractions that we found between

---

[3]We recall that whole numbers (like 1) are fractions just like any other fraction. A whole number is simply a fraction whose "bottom" number is just 1. As examples, we have that $1 = \frac{1}{1}$, $5 = \frac{5}{1}$, and $17 = \frac{17}{1}$. This is because if we have 17 pizzas and only one person to distribute them to, then that person gets 17 pizzas.

[4]We also note that there are lots of **other** fractions between $\frac{1}{2}$ and 1 that we never obtained throughout this "chopping in half" procedure. For example, $\frac{2}{3}$ is between $\frac{1}{2}$ and 1, as is $\frac{3}{5}$, as is $\frac{9}{7}$, yet these (and infinitely many other fractions between $\frac{1}{2}$ and 1) were never "hit" by our procedure of chopping things in half.

$\frac{1}{2}$ and 1 was the fraction $\frac{3}{4}$. We found this fraction by taking a step halfway between $\frac{1}{2}$ and 1. However, we then immediately know that there are infinitely many fractions between $\frac{1}{2}$ and $\frac{3}{4}$. In particular, $\frac{5}{8}$ is halfway between $\frac{1}{2}$ and $\frac{3}{4}$, and then $\frac{11}{16}$ is halfway between $\frac{5}{8}$ and $\frac{3}{4}$, and then $\frac{23}{32}$ is halfway between $\frac{11}{16}$ and $\frac{3}{4}$. We can cut this interval between $\frac{1}{2}$ and $\frac{3}{4}$ in half as many times as we want, thus getting yet another infinity of fractions that are all still between $\frac{1}{2}$ and 1, but which we did not see when cutting the interval between $\frac{1}{2}$ and 1 in half over and over. And in just the same way, we can pick any two successive fractions that we just found — for example, we can take the fractions $\frac{11}{16}$ and $\frac{23}{32}$ — and cut **that** interval in half over and over again, thus finding **yet another** new infinity of fractions.

In short, every time we take a step halfway from one fraction to another,[5] we can find another infinity of fractions between the two we just stepped between. And then, between any two fractions that we step between, we will be skipping yet **another** new infinity of fractions. We can then continue this on and on, looking between any two fractions and finding an infinity of fractions between them, and then an infinity of fractions between any of the two fractions in this infinity of fractions, and then an infinity of fractions between any of the two fractions in this infinity of fractions, and so on, forever.

This sort of infinity certainly has a different quality about it than the "infinity type 1" of $\{1, 2, 3, ...\}$. Namely, $\{1, 2, 3, ...\}$ just kind of "goes in one direction," so that whenever we look between two elements in this set we will always only find finitely many elements. However, in the set $FRAC$ of all fractions, we can "zoom in" on the fractions infinitely far, in infinitely many places, infinitely many times, and still continue to find more fractions! This is what we mean by "infinity times infinity." Between any two of the infinitely many elements in FRAC, there are infinitely many elements, and between any two of those elements, there are infinitely many elements, and between any two of those elements, there are infinitely many elements, and on and on and on!

Because of this fundamental difference in the qualitative nature of the set of positive whole numbers $\{1, 2, 3, ...\}$ and the set $FRAC$,

---

[5] For absolutely any two fractions that we choose.

A SEEMINGLY NEW INFINITY

we may be tempted to think that there cannot possibly be a bijective function from the former set to the latter set — from $\{1, 2, 3, ...\}$ to $FRAC$. To see that we can define an injective function is easy, since every whole number is itself a fraction. Thus we could define an injective function just by sending 1 to 1, and 2 to 2, and 3 to 3, and so on. But this function certainly is not surjective. In fact, it might seem pretty unlikely that we can "hit" all of the fractions just with $\{1, 2, 3, ...\}$ since, for example, no matter where we send 1 and 2 we will be skipping over infinitely many fractions. Believe it or not, however, we actually **still can** define a bijective function from $\{1, 2, 3, ...\}$ to $FRAC$. It will take some work and some creativity to see this, and so this will be the subject of the entire next chapter. Then, after using Chapter 13 to develop some more machinery for proofs in general, we will see a whole new kind of infinity in Chapter 14!

Chapter 12

Infinity Times Infinity = Infinity

Needing To Get Creative

In the previous chapter we saw how the set of all fractions (which we denoted by $FRAC$) is **extremely** infinite. In particular, we had "infinitely many infinities" due to the fact that between any two fractions there are infinitely many fractions, and between any of those fractions there are again infinitely many fractions, and so on, forever. Moreover, the initial two fractions that we chose in this process could have been any of the infinitely many fractions that exist! In this sense, $FRAC$ appears to have "infinity times infinity" elements. Of course, the notion of "infinity times infinity" is not yet well-defined, because the only infinity that we have defined is "infinity type 1," which is the infinity corresponding to the set $\{1, 2, 3, 4, ...\}$.

This chapter will be devoted to showing that this vague idea of "infinity times infinity" is in fact the same as "infinity type 1." In other words, we will find a bijective function from $\{1, 2, 3, 4, ...\}$ to $FRAC$, showing that the infinity of the fractions is actually the same as the infinity of $\{1, 2, 3, ...\}$. It may seem surprising that this is possible since we have seen how the "infinite-ness" of $FRAC$ is so seemingly different from the "infinite-ness" of $\{1, 2, 3, 4, ...\}$. In particular, the set $\{1, 2, 3, ...\}$ does **not** have the property that between

NEEDING TO GET CREATIVE

any two of its elements there are infinitely many more elements. And since $FRAC$ **does** have this property, we will have to get a lot more creative in defining this bijective function than we did when we defined a bijective function from $\{1, 2, 3, ...\}$ to $\{1, -1, 2, -2, 3, -3, ...\}$ in Chapter 10.

The reason why we have to get more creative now is relatively easy to see. For the function from $\mathbb{N} = \{1, 2, 3, ...\}$ to $B = \{1, -1, 2, -2, 3, -3, ...\}$ in Chapter 10, we were able to exploit the fact that for any positive number there is an obvious "next" positive number,[1] and that for any negative number there is an obvious "next" negative number. Namely, the "next" positive number after 2 is 3, and the "next" one after 3 is 4, and so on. Similarly, the "next" negative number after -2 is -3, and the "next" negative number after -3 is -4, and so on. Similarly, for any positive even number there is an obvious "next" positive even number, and for any positive odd number there is an obvious "next" positive odd number. For example, the "next" even number after 2 is 4, and the "next" even number after 4 is 6, and so on. Similarly, the "next" odd number after 3 is 5, and the "next" odd number after 5 is 7, and so on. We were then able to send the odd numbers in $\mathbb{N}$ to the positive numbers in B, and the even numbers in $\mathbb{N}$ to the negative numbers in B. Moreover, we were able to do so without missing any numbers in B since we were able to send the "next" odd number in A to the "next" positive number in B, and the "next" even number in $\mathbb{N}$ to the "next" negative number in B.

The problem with $FRAC$ is that for any fraction, there is no meaningful notion of the "next" fraction. To be clear, it most certainly **is** the case that if the enemy handed us two distinct fractions then we are of course able to say which one is bigger. However, suppose the enemy handed us the fraction $\frac{1}{2}$ and asked us what the "next biggest" fraction is. We would quickly respond by saying that the enemy's question is impossible to answer. This is because, as we have seen, if

---

[1] As usual, scare quotes are used here only because the notion of "next" positive number has not yet been rigorously defined. Thus, even though it is obvious what the "next" positive number after 2 is, we must still remember that we do not yet actually know what "next" means. Indeed, in a later volume we will make the notion of "next" mathematically precise, but for now we will simply use our intuitive understanding of the word.

we were to choose any fraction at all as the "next" fraction, then we will have immediately skipped over infinitely many fractions between $\frac{1}{2}$ and whichever fraction we chose. It is for this reason that for any fraction at all, there simply does not exist any notion of the "next biggest" fraction. We can therefore see that there will be no nice and obvious function between these two sets, and that is why we need to get creative (which is not the end of the world).

Finding The Function

Let us now construct this function. We will in fact only construct a function from the **even** numbers in $\mathbb{N} = \{1, 2, 3, ...\}$ to the **positive** elements of $FRAC$ (recall that fractions can be positive **or** negative). Once we have done this it will be obvious that we can finish the job by sending the **odd** elements of $\mathbb{N}$ to the **negative** elements of $FRAC$ in the exact same way. We begin by recalling the following two facts. Firstly, we recall that every fraction is nothing but "some whole number upstairs, and some (non-zero) whole number downstairs." In other words, every fraction is nothing but a numerator and a denominator — both of which are whole numbers, and the denominator is not zero. Secondly, we recall that $\frac{1}{2}$ and $\frac{2}{4}$ and $\frac{3}{6}$ and $\frac{4}{8}$ (etc.) are all the same element in $FRAC$. More generally, we recall that two fractions are equal if one can be obtained from the other by multiplying the numerator and the denominator by the same thing.

Using the first fact as motivation, let us construct an infinitely large chart where positive whole numbers go across the top and down the left. This is depicted in Figure 12.1, where the positive whole numbers across the top are the numerators and those going down the left are the denominators. For example, the fraction in the box with numerator 7 and denominator 1 is the fraction 7, and the fraction in the box with numerator 8 and denominator 9 is $\frac{8}{9}$. Note that this chart goes infinitely far in both directions since it includes **all** of the positive whole numbers on **each** axis. Clearly Figure 12.1 only shows a small portion of the entire infinite chart, primarily because it would take us an infinite amount of time[2] to show the whole thing!

---

[2]And it would require an infinitely large page.

FINDING THE FUNCTION

Numerators→ Denominators↓	1	2	3	4	5	6	7	8	9	10	...
1	1	2	3	4	5	6	7	8	9	10	...
2	$\frac{1}{2}$	$\frac{2}{2}$	$\frac{3}{2}$	$\frac{4}{2}$	$\frac{5}{2}$	$\frac{6}{2}$	$\frac{7}{2}$	$\frac{8}{2}$	$\frac{9}{2}$	$\frac{10}{2}$	...
3	$\frac{1}{3}$	$\frac{2}{3}$	$\frac{3}{3}$	$\frac{4}{3}$	$\frac{5}{3}$	$\frac{6}{3}$	$\frac{7}{3}$	$\frac{8}{3}$	$\frac{9}{3}$	$\frac{10}{3}$	...
4	$\frac{1}{4}$	$\frac{2}{4}$	$\frac{3}{4}$	$\frac{4}{4}$	$\frac{5}{4}$	$\frac{6}{4}$	$\frac{7}{4}$	$\frac{8}{4}$	$\frac{9}{4}$	$\frac{10}{4}$	...
5	$\frac{1}{5}$	$\frac{2}{5}$	$\frac{3}{5}$	$\frac{4}{5}$	$\frac{5}{5}$	$\frac{6}{5}$	$\frac{7}{5}$	$\frac{8}{5}$	$\frac{9}{5}$	$\frac{10}{5}$	...
6	$\frac{1}{6}$	$\frac{2}{6}$	$\frac{3}{6}$	$\frac{4}{6}$	$\frac{5}{6}$	$\frac{6}{6}$	$\frac{7}{6}$	$\frac{8}{6}$	$\frac{9}{6}$	$\frac{10}{6}$	...
7	$\frac{1}{7}$	$\frac{2}{7}$	$\frac{3}{7}$	$\frac{4}{7}$	$\frac{5}{7}$	$\frac{6}{7}$	$\frac{7}{7}$	$\frac{8}{7}$	$\frac{9}{7}$	$\frac{10}{7}$	...
8	$\frac{1}{8}$	$\frac{2}{8}$	$\frac{3}{8}$	$\frac{4}{8}$	$\frac{5}{8}$	$\frac{6}{8}$	$\frac{7}{8}$	$\frac{8}{8}$	$\frac{9}{8}$	$\frac{10}{8}$	...
9	$\frac{1}{9}$	$\frac{2}{9}$	$\frac{3}{9}$	$\frac{4}{9}$	$\frac{5}{9}$	$\frac{6}{9}$	$\frac{7}{9}$	$\frac{8}{9}$	$\frac{9}{9}$	$\frac{10}{9}$	...
10	$\frac{1}{10}$	$\frac{2}{10}$	$\frac{3}{10}$	$\frac{4}{10}$	$\frac{5}{10}$	$\frac{6}{10}$	$\frac{7}{10}$	$\frac{8}{10}$	$\frac{9}{10}$	$\frac{10}{10}$	...
⋮	⋮	⋮	⋮	⋮	⋮	⋮	⋮	⋮	⋮	⋮	⋱

Figure 12.1: Infinite in both directions in order to include all the positive fractions (with many duplicates).

The important thing to note about this chart is that it includes **every** positive fraction. This is because it contains every possible combination of numerator and denominator, as long as we limit ourselves to the positive fractions. We also note that there are (infinitely) many duplicate fractions in the chart. For example, all of the fractions along the diagonal of the chart — where the numerators and the denominators are equal to each other — are equal to one.[3] There are also lots of other duplicates since, for example, $\frac{1}{2}=\frac{2}{4}=\frac{3}{6}$ and $\frac{3}{4}=\frac{6}{8}$, and so on. The fact that there are (lots of) duplicates in Figure 12.1 will not be a problem for us — the only important thing is that every positive fraction is contained **somewhere** on this infinite chart.

All we need to do now is take the even elements of $\mathbb{N} = \{1, 2, 3, ...\}$ and assign them in a clever way to this chart. In particular, what we need to do is assign the even numbers $\{2, 4, 6, ...\}$ to this chart in such a way that we hit every single square. By doing so, we will have defined a **surjective** function from the even numbers $\{2, 4, 6, ...\}$ to

---

[3]This is simply because $\frac{1}{1}=\frac{2}{2}=\frac{3}{3}=\frac{4}{4}=\cdots=1$.

the positive fractions. Then all we need to do is make this function **injective** by "skipping" over the duplicate fractions, so that we do not hit the same element[4] twice.

The idea will be to "walk along" this chart, placing the next even number down on the next square (as long as this square is not a duplicate of a square that we have already walked over). But we need to be careful about how we walk around this chart, since it is infinitely large. Namely, we cannot say that we will start with the first row and then move on to the second row, because the first row is infinitely long! Thus, if we started walking down the first row placing subsequent even numbers on subsequent squares, then we will never get out of the first row! Therefore, what we really need to do is "zig-zag" our way through this chart in such a way that every (non-duplicate) square is eventually hit. Let us see how we can do this.

We begin by assigning 2 in $\mathbb{N}$ to the top-left-most element in the chart, which is the fraction 1. This is depicted in Figure 12.2 by placing a 2 in the top-left corner of the box with the fraction 1 in it. Thus, the numbers in the top-left corner of each square correspond to the elements in the domain (i.e., $\mathbb{N}$) of our function from $\mathbb{N}$ to $FRAC$, and the bigger numbers in the center of each square correspond to the fractions that our function sends those elements to. The idea now is to head to the next box — whatever we decide the next box to be — and we place the next even number there as long as the fraction in this box has not already been hit by our function. Let us now take a step to the right, which is depicted in Figure 12.2 by an arrow from the box with the fraction 1 in it to the box with the fraction 2 in it. We are now sitting on the square containing the fraction 2, and our function has **not** yet hit the fraction 2, so we assign the next even number, which is 4, to this square.

Since we know that we do not want to keep heading down the same row (or else we will be forever stuck in this one row), we now head diagonally down and to the left. This is depicted in Figure 12.2 by an arrow pointing from the box containing the fraction 2 to the box containing the fraction $\frac{1}{2}$. Since our function has yet to hit the

---

[4]In this case, the same fraction.

FINDING THE FUNCTION

fraction $\frac{1}{2}$, we assign the next even number, which is 6, to this square. We now take a step down, so that we are standing on the square with $\frac{1}{3}$ in it. Our function has not yet hit the fraction $\frac{1}{3}$ and so we assign the next even number, which is 8, to this square.

Numerators → Denominators ↓	1	2	3	4	5	6	7	8	9	10	...
1	1	2	3	4	5	6	7	8	9	10	...
2	$\frac{1}{2}$	$\frac{2}{2}$	$\frac{3}{2}$	$\frac{4}{2}$	$\frac{5}{2}$	$\frac{6}{2}$	$\frac{7}{2}$	$\frac{8}{2}$	$\frac{9}{2}$	$\frac{10}{2}$	...
3	$\frac{1}{3}$	$\frac{2}{3}$	$\frac{3}{3}$	$\frac{4}{3}$	$\frac{5}{3}$	$\frac{6}{3}$	$\frac{7}{3}$	$\frac{8}{3}$	$\frac{9}{3}$	$\frac{10}{3}$	...
4	$\frac{1}{4}$	$\frac{2}{4}$	$\frac{3}{4}$	$\frac{4}{4}$	$\frac{5}{4}$	$\frac{6}{4}$	$\frac{7}{4}$	$\frac{8}{4}$	$\frac{9}{4}$	$\frac{10}{4}$	...
5	$\frac{1}{5}$	$\frac{2}{5}$	$\frac{3}{5}$	$\frac{4}{5}$	$\frac{5}{5}$	$\frac{6}{5}$	$\frac{7}{5}$	$\frac{8}{5}$	$\frac{9}{5}$	$\frac{10}{5}$	...
6	$\frac{1}{6}$	$\frac{2}{6}$	$\frac{3}{6}$	$\frac{4}{6}$	$\frac{5}{6}$	$\frac{6}{6}$	$\frac{7}{6}$	$\frac{8}{6}$	$\frac{9}{6}$	$\frac{10}{6}$	...
7	$\frac{1}{7}$	$\frac{2}{7}$	$\frac{3}{7}$	$\frac{4}{7}$	$\frac{5}{7}$	$\frac{6}{7}$	$\frac{7}{7}$	$\frac{8}{7}$	$\frac{9}{7}$	$\frac{10}{7}$	...
8	$\frac{1}{8}$	$\frac{2}{8}$	$\frac{3}{8}$	$\frac{4}{8}$	$\frac{5}{8}$	$\frac{6}{8}$	$\frac{7}{8}$	$\frac{8}{8}$	$\frac{9}{8}$	$\frac{10}{8}$	...
9	$\frac{1}{9}$	$\frac{2}{9}$	$\frac{3}{9}$	$\frac{4}{9}$	$\frac{5}{9}$	$\frac{6}{9}$	$\frac{7}{9}$	$\frac{8}{9}$	$\frac{9}{9}$	$\frac{10}{9}$	...
10	$\frac{1}{10}$	$\frac{2}{10}$	$\frac{3}{10}$	$\frac{4}{10}$	$\frac{5}{10}$	$\frac{6}{10}$	$\frac{7}{10}$	$\frac{8}{10}$	$\frac{9}{10}$	$\frac{10}{10}$	...
⋮	⋮	⋮	⋮	⋮	⋮	⋮	⋮	⋮	⋮	⋮	⋱

Figure 12.2: The first several assignments in our function from the even elements of N to the positive elements of *FRAC*. The even number assigned to any given box is shown by putting an even number in the top-left corner of that box. The boxes without numbers in the top-left corners have X's because those boxes are duplicates of fractions that our function has hit before, and so our function skips over them. These assignments cover the entire infinite grid, but we only show the assignments for the even numbers up to 40 because we do not have the infinite time required to show all of the infinitely many assignments.

We now head diagonally back up and to the right, so that we are standing on the square with $\frac{2}{2}$ in it. However, $\frac{2}{2} = 1$ and we have already assigned an even number to 1 (it was in fact our first assignment). Let us therefore put an "X" in this square and skip it — i.e., let us **not** assign an even number to this square — and continue to head up and to the right again. We are now standing on the square containing 3. Since we have not assigned any even number to 3 yet, we assign the next even number, which is 10, to 3.

We continue this process by moving one square to the right and then moving diagonally down and to the left until we are at the left edge of the chart. We then move one square down and then diagonally up and to the right until we are back at the top edge of the chart. At each step we assign the next even number if the box that we are standing on is not a duplicate of any previous box that we have been on (while putting an "X" on and skipping the box otherwise). We are therefore systematically "zig-zagging" back and forth through this chart so that we never get stuck in any of the infinitely many rows or columns. Thus, since we have infinitely many even numbers to use, we will eventually hit **every single** positive fraction. This is because every single positive fraction is **somewhere** on this chart, and since our "zig-zagging" method allows us to not miss any squares, we will eventually arrive at **any** fraction that the enemy chooses. To illustrate how we "zig-zag" through the chart, we have shown how to assign[5] the even numbers up to 40.

Of course, these assignments continue on infinitely far. The main point, however, is that if the enemy chose any positive fraction at all, then there is some (possibly very large) even number that eventually hits it. This is the definition of surjectivity — namely, we do not miss any fractions. Additionally, this function from the even numbers to the positive fractions is injective due to our requirement of skipping duplicate fractions (so that no two even numbers are sent to the same fraction). Thus, our function is bijective!

This is, however, only a function from the **even** numbers to the **positive** fractions. To finish the map from $\mathbb{N}$ to $FRAC$, we simply

---

[5]As well as how to skip. Namely, we see that every square on the diagonal after the top-left-most square will have an "X" on it because all of those squares equal 1. Similarly, the square containing the fraction $\frac{2}{4}$ is skipped because it is equal to $\frac{1}{2}$, which we have already hit. As another example, the square containing the fraction $\frac{2}{6}$ is also skipped because it is equal to the fraction $\frac{1}{3}$, which we have also already hit.

FINDING THE FUNCTION

need to make **another** chart with all of the **negative** fractions and assign the **odd** elements of ℕ to these fractions in the exact same way that we have for the positive fractions and even elements of ℕ. Namely, by reproducing the entire argument of this section but with the word "odd" replacing "even" and "negative" replacing "positive," we can hit the rest of the fractions (and do so bijectively) using the odd elements of ℕ.

Finally, we note that we still need to assign a number in ℕ to the fraction 0 because 0 has not made an appearance in our discussion so far. Namely, the fraction 0 is not in the infinitely large chart of Figure 12.1 because 0 is not positive. It also would not have been in the second chart of negative fractions that we just described (but did not explicitly display) because 0 is not negative. We know, though, that we can add any finite number of things to a set and still not ruin that set's type of infinity. Thus, all we need to do is pick either the even numbers or the odd numbers in ℕ and simply "shift" them up by 1. For example, instead of choosing 2 as the first even number that we use in the chart in Figure 12.2, we could use 4 as the first even number and shift all the remaining even numbers up to the next even number. Then, we simply assign the even number 2 to 0. We therefore see that once we have covered the positive fractions with the even numbers and the negative fractions with the odd numbers, including 0 is not a problem and we are therefore done showing that there exists a bijective function from ℕ to $FRAC$.

And with that, we are done! We have successfully tamed the wild infinities that exist in $FRAC$ and shown that $FRAC$ is really "the same infinity" as $\{1, 2, 3, ...\}$. In particular, despite our initial thoughts to the contrary, we have found that the set $FRAC$ and the set $\mathbb{N} = \{1, 2, 3, ...\}$ both have exactly "infinity type 1" elements in them since we have found a bijective function from ℕ to $FRAC$.

Let us quickly note that the function that we have constructed here is in no way "nice," meaning that it does not have a nice pattern to it. Let us denote by F this function[6] from ℕ to $FRAC$, so that

---

[6] Let us, for the sake of the next sentence, suppose that we have removed the fraction 0 from the set $FRAC$. As we have seen, removing one element from an infinite set will not effect that set's cardinality, and if we temporarily forget about "hitting" the fraction 0 then what follows will align correctly with Figure 12.2.

we have $F : \mathbb{N} \to FRAC$. We then see by examining Figure 12.2 that $F(2) = 1$, $F(4) = 2$, $F(6) = \frac{1}{2}$, $F(8) = \frac{1}{3}$, and $F(10) = 3$, for example. Similarly, by constructing the second chart of negative fractions and assigning the odd numbers to them, we see that $F(1) = -1$, $F(3) = -2$, and $F(5) = -\frac{1}{2}$, for example. The farther into the chart we go, the more this function "jumps around." But it does not matter! The function is most definitely surjective and injective, and so it is bijective, and thus with our definitions we are **forced** to say that $\{1, 2, 3, ...\}$ is "just as infinite" as $FRAC$.

Seeing that a set as "wildly infinite" as $FRAC$ also only has "infinity type 1" elements may make us begin to think that **all** infinite sets have "infinity type 1" elements. However, we **will** see a brand new type of infinity — a much larger infinity — in Chapter 14. First, however, we must take some time to get used to the powerful logic behind the proof that we will require in that chapter. This method of proof goes by the name "Proof By Contradiction" and we will spend the next chapter seeing what it is all about. And once this method is at our fingertips, it will allow us to sail off into the amazing world of the infinite, where infinitely many increasingly large infinities rise up endlessly, towering over all of the infinities that come before them.

Chapter 13

Proof By Contradiction

Axioms

Before continuing to dive deeper into the whacky world of the infinite, we need to pause for a moment and introduce a new tool into our mathematical toolbox. As mathematicians we must continually develop the tools and/or weapons that we need in order to be able to continue to access increasingly complex, subtle, and beautiful truths. In this regard, the method of proof that we will develop here — namely, the method that goes by the name "proof by contradiction" — is important not only for accessing the truth that we will uncover in the next chapter, but also for our development as mathematicians as a whole.

The basic idea behind the method of proof by contradiction is the following. One way to prove that a given statement is true is to **suppose** that it is false. If, upon assuming that this statement is false, we can derive some kind of ridiculous statement, then we will in fact have proven that the statement is true. Namely, if we can use the **supposed** falseness of this statement to arrive at a **contradiction**,[1] then we will know that the statement must in fact **not** be false, and must therefore be true. Similarly, if we want to prove that some statement is **false**, then we can **suppose** that it is true and then arrive at a contradiction. In this case, since math should not have

---

[1] Namely, a statement that is simultaneously true and false.

any contradictions, we are then forced to accept that the statement is false.

We note that we use this type of logic regularly even in non-mathematical settings. For example, suppose we forgot whether or not we fed the dogs earlier this evening. Perhaps we have a suspicion that we **did** feed the dogs, but wanted to find some reassurance in order to know that we are not depriving our dogs of their dinner. We walk over to where the dogs' dishes are and see that indeed there are two empty dishes sitting there. Suppose that we distinctly remember seeing the dishes in the garage earlier this morning, which is where we put them after the dogs were done eating the previous night. We can then be sure that we did indeed feed the dogs this evening. This is because if we suppose we did **not** feed the dogs, then the dishes were never moved from the garage. However, the dishes are very clearly in the kitchen (where the dogs always eat). These are two contradictory facts and therefore it must be that our **supposition** that we did not feed the dogs was false.

This may all seem like a somewhat unnecessary analysis of a simple scenario, but this type of logic is extremely important[2] in math. Therefore, let us turn our attention towards how it applies to the more well-defined statements of mathematics, as opposed to just the statements about caring for one's house pets.

We recall that mathematics is about making precise definitions of mathematical structures and using irrefutable logic to deduce new truths about those structures from old ones. Underneath the hood of any mathematical field is some set of **axioms**, which are statements that are taken to be true without explanation or proof. For example, our "One True Sentence" from the first section of the first chapter of this volume is a mathematical axiom. We cannot **derive** the fact that "something exists" from more fundamental principles, but rather we are **supposing** that something exists.[3] Once we agree that this statement **must** be true, then we can define elements and sets and start building up mathematics by proving various things about these

---

[2] And even somewhat controversial in some mathematical circles, but for reasons that we will not see for a very long time.

[3] Otherwise, if nothing existed, we would have much larger issues on our hands than that of developing mathematics.

elements and sets.

Any subfield of math has its own set of axioms. For example, we have had to take as an **axiom** the fact that numbers exist,[4] that there are infinitely many numbers, that there are only even and odd numbers, and so on.[5] Once we posited these truths as axioms we were able to define concepts like cardinality and "infinity type 1," and then we were able to start proving things about those concepts.

We quickly note that in any mathematical field, the **fewer** the axioms that are required the better. This is simply because any mathematical truth is more logically sound (and therefore more beautiful) when it can be derived from a smaller set of assumptions.[6] If some set of mathematical truths can be derived from 1, 2, or 3 axioms, then those truths are somehow more logically satisfying than they would be if we had to make hundreds or thousands of axioms before being able to arrive at their truth. Of course, **anything** could be **technically** true by simply making it an axiom. We could come up with our own field of math whose foundational axiom is that pigs can fly, though the lack of mathematical precision and the questionable truth of this statement would probably result in our theory not gaining much traction within the mathematical community.

Statements And Contradictions

We can now see that any mathematical theory is essentially constructed as follows.[7] There is some collection of axioms that we take

---

[4] There are indeed mathematicians who try to understand how the existence of numbers might indeed be derivable from some more fundamental concepts, but we will leave these issues aside here as they will take us too far away from the main thread of mathematics that we want to explore.

[5] Some of these concepts are indeed **not** axioms, but rather can be proven from fundamental principles. Again, though, we will leave these issues aside because they are not important for us.

[6] The word "assumption" here is perhaps too weak, because mathematical axioms are hardly "assumptions" in the sense that we usually think of the word — a mathematical axiom is an assumption whose truth is so blatantly obvious that no one would refute its validity. For example, not too many would argue statements like "numbers exist."

[7] This is also a bit subtle, and the reasons for this subtlety are fascinating, but we will not see them for several more volumes.

to be true without explanation, and then there are the statements that we derive **from** these axioms using nothing but the truth of the axioms and logical deduction. In this way, once the axioms are taken to be true, so too must all of the statements that are (correctly) derived from those axioms, otherwise we would be left in the intolerable position of having true statements and correct reasoning lead to incorrect statements.

We can therefore visualize the mathematical process as follows. Suppose we have some (ideally small) collection of mathematical axioms. We may as well give these axioms names, and call them A_1, A_2, A_3, ... A_N, where N is simply the total number of axioms.[8] Suppose also that we decide to make a list of all mathematical facts that we know — with certainty — to be true. Before we start "doing math" — i.e., before we start proving any new statements — the only facts that will be in our list of true facts will be the axioms (simply because these are assumed to be true even before we start doing any work). Now suppose that we (or some other mathematicians) took our axioms and, using nothing but logical deductions, proved the validity of some new statement. Let us give this statement a name and call it S_1. In other words, some mathematicians somewhere (preferably us) did some work — they thought about the axioms for a long time — and were able to show that if A_1 and A_2 and A_3 and ... and A_N are all true, then it **must** be the case that S_1 is true also.[9]

This is significant progress, because now our collection of true statements has expanded from $\{A_1, A_2, A_3, ..., A_N\}$, (i.e., just the axioms) to $\{A_1, A_2, A_3, ..., A_N, S_1\}$. We can visualize this as

$$\{A_1, A_2, A_3, ..., A_N\} \xrightarrow{\text{Blood, Sweat, and Tears}} \{A_1, A_2, A_3, ..., A_N, S_1\},$$

where the "Blood, Sweat, and Tears" is that of hard working mathematicians like ourselves. We have already seen examples of this. In

---

[8] As mentioned before, it is ideal that N is a small number, but for now we can just let it be any number.

[9] Recall that A_1, A_2, A_3, ..., A_N, and S_1 here are all symbols denoting **entire statements** so that, for example, we might have $S_1 = $ "$1 + 1 = 2$" or $S_1 = $ "Kobe is better than LeBron."

particular, S_1 might have been "the even numbers have the same cardinality as $\mathbb{N}$," since this was a true statement that we derived after some thought, shedding some blood, sweat, and tears, and using the knowledge of the true facts that we had already established.

After we have established the validity of this new statement S_1, we can then go on to try to establish the validity of more statements. However, we are now able to **use** S_1 when deriving some new statement S_2, since we have already established that S_1 is true. Namely, we do not have to keep deriving all of our new statements from the foundational axioms, but rather we can continue to use the new statements that we have **already** established to be true in order to uncover new truths. We can therefore visualize the abstract progression of mathematics as a growing list of true statements, where at each stage all of the known truths can be derived from the truths of the previous stage (until we get back to the original set of axioms, at which point the train stops). We then have a logical flow of collections of true statements that can be depicted as

$$\{A_1, A_2, A_3, ..., A_N\} \xrightarrow{\text{Effort}} \{A_1, A_2, A_3, ..., A_N, S_1\} \xrightarrow{\text{Effort}}$$
$$\{A_1, A_2, A_3, ..., A_N, S_1, S_2\} \xrightarrow{\text{Effort}} \{A_1, A_2, A_3, ..., A_N, S_1, S_2, S_3\}$$
$$\xrightarrow{\text{Effort}} \{A_1, A_2, A_3, ..., A_N, S_1, S_2, S_3, S_4\} \xrightarrow{\text{Effort}} \cdots.$$

We note that there are many different ways to uncover new truths from previously established truths, and this corresponds to the fact that there are many different logical routes to the same truths. Thus, the "Effort" stage of this process is not necessarily direct — it usually takes a lot of creativity, ingenuity, and insight. For example, the "Effort" stage was much more involved for proving the statement "the set of fractions has the same cardinality as the set $\mathbb{N}$" than the "Effort" stage was for proving the statement "the set of even numbers has the same cardinality as the set $\mathbb{N}$." The method known as "proof by contradiction" is one very powerful method for establishing new truths from some collection of previously established truths. Indeed, proof by contradiction is not only extremely powerful, but it is also extremely **general**, in the sense that it can be used in virtually every imaginable field of mathematics as a valid way to take some pre-established truths and derive new truths. Let us therefore see how this method works.

Suppose at some stage in the development of some mathematical framework we have found ourselves in the situation where the statements $\{A_1, ..., A_N, S_1, ..., S_M\}$ are already established to be true. Namely, suppose that we are dealing with N axioms and that we have already established M additional and completely irrefutable truths from these axioms. Now suppose we want to ask whether or not some statement P is true. The statement P can be any well-defined mathematical statement, where by "well-defined" we simply mean that P is some statement that has **either** a true **or** a false outcome — not both, and not neither.

There are several different ways that we might be able to determine if P is true (and can therefore be added to our collection $\{A_1, ..., A_N, S_1, ..., S_M\}$ of true statements) or false (and therefore **cannot** be added to our collection of true statements). For example, we could try to prove it directly using only the axioms and the previously established true statements, in which case we just have to be sufficiently clever and creative. However, we could also do the following. Suppose we have a sneaking suspicion that P is, for example, **not** true. Our sneaking suspicion would of course only be confirmed if we can actually prove that P is not true, but we can let our suspicion guide what we do. Namely, if we want to prove that P is **not** true, we can first **suppose** that it **is** true. Namely, we can pretend (for the moment) that P is true and include it in our list temporarily. Thus, we can consider what happens if we had the list $\{A_1, ..., A_N, S_1, ..., S_M, P\}$.

Once we make this supposition we can then start proving new true statements from our new (extended) list $\{A_1, ..., A_N, S_1, ..., S_M, P\}$ of "true" statements.[10] If we can use this new set of "true" statements to prove that one of the **previously proven** statements is **false**, then we will have arrived at a contradiction. Namely, suppose that we can use this new list $\{A_1, ..., A_N, S_1, ..., S_M, P\}$ of true statements to prove that one of the statements A_1, A_2, ..., A_N, S_1, S_2, ..., or

---

[10] We use scare quotes here because we must remember that the truth of P is still in question.

S_M is false. Then our logical deductions will have shown that some statement is **both** true and false, which is clearly absurd. In particular, it should never be the case that **correctly adding** a new true statement to our list makes a **previously** proven statement no longer true. This is because each successive true statement that we derive relies on the truth of the statements that come before it. Thus, if we add a statement to our list that genuinely belongs in the list (i.e., is **actually** true) then it must not call into question the truth of any of the statements preceding it. Otherwise the truth of the preceding statements will have led to their own falseness — a clear absurdity!

To summarize, we use the method of contradiction to prove the truth (or falseness) of a given statement as follows. If we want to prove that a certain statement is true, then we can **suppose** that it is false. We then derive more and more statements **from the assumption that this statement is false**, and we try to prove the falseness of something that we already knew to be true (prior to the assumption that the statement we are inquiring about is false). This then shows that we **incorrectly** assumed that the statement we started with is false, and therefore it must be true. Similarly, if the statement we started with is indeed false and we want to prove that it is so, then we can **suppose** that it is true. We then must use the supposed truth of this statement to find a contradiction with some other previously proven statement, and by doing so we show that our supposition was wrong and that our statement is indeed false. We will now see a very famous and very beautiful example of this method of proof, and it is provided by Euclid's proof of the fact that there are infinitely many prime numbers.

Example: Infinitely Many Primes

Before seeing Euclid's clever proof of the fact that there are infinitely many prime numbers, there are a few basic facts that we need to recall. First, we need to recall what a prime number is in the first place. By definition, a prime number is a number that cannot be divided by any smaller number. For example, 6 is **not** a prime number because 6 can be divided by 3 (to give 2), as well as by 2 (to give 3). Similarly, 10 is **not** a prime number because it can be divided by 2 (to give 5) and 36 is **not** a prime number because it can be divided by 6 (amongst other things). However, 3 **is** a prime number because it can only be divided[11] by 3 and by 1. Similarly, 2 is a prime number, 5 is a prime number, 7 is a prime number, and 11 is a prime number.

The final stipulation on prime numbers is that they are positive and that they start at 2. Namely, 2 is the smallest prime number. We note that so far there is no reason why 1 would not be a prime number, since after all it is a number that is only divisible by itself and by 1. However, this is simply a definition, and we can define prime numbers however we want. For reasons that we will see in the future, the mathematical community has chosen to define prime numbers to start at 2. We therefore have the following definition.

Definition 13.1. A prime number is an element of the set $\mathbb{N}$ that is greater than or equal to 2, and which is only divisible by itself and by 1.

One thing we can note immediately is that 2 is the only prime number that is also even. This is easily seen by noting that any even number that is **larger** than two cannot be prime since it would be divisible by 2 (by the very definition of an even number). Let us, for the sake of gaining some familiarity with prime numbers, explicitly write out the first fifteen prime numbers:

$$\{2, 3, 5, 7, 11, 13, 17, 19, 23, 29, 31, 37, 41, 43, 47\}.$$

Now that we have made a definition (namely, that of a prime number), we can go on to ask questions about it. We were free to make our definition however we want, but now we are going to be stuck with its consequences. The question that we will ask here is whether or not there are infinitely many prime numbers. Namely, let us see if we can figure out if the prime numbers "go on forever," or instead if there is some largest prime number, after which every single number is no longer prime.

It is important to take a second to reflect on the fact that the way in which we might answer this question is not clear. For suppose

---

[11]We note that **any** number can be divided by itself (to give 1) and by one 1 (to give itself). Prime numbers are those numbers for which these are the **only** two options.

EXAMPLE: INFINITELY MANY PRIMES

we continued the above list of prime numbers until we got to the $1,000^{th}$ prime number. We might then be tempted to say that the prime numbers go on forever, but we would be wrong to make this assertion. This is because we still would not know if it were the case that, for example, the $1,000,000^{th}$ prime number is the largest prime number and then the prime numbers stop after that. Conversely, suppose we wrote down the first million prime numbers and then found that none of the next ten billion numbers were prime. We might then be tempted to say that the $1,000,000^{th}$ prime number is indeed the largest prime number and that therefore there are not infinitely many prime numbers. However, it very well could be the case that even though the next ten billion numbers are not prime, perhaps some number that is a hundred billion billion billion times greater than the $1,000,000^{th}$ prime number is another prime number. Thus, there is no "direct" way to answer this question, in the sense that there is no number that we can just "compute" in order to answer the question, and we therefore see that we need to get more creative in order to find the answer.

Let us also reflect on the fact that the answer itself is not all that clear. Namely, it is sometimes the case that the answer to a question might be obvious while the **proof** of that answer is not. In the preceding paragraph we saw that the proof of whatever the answer is will not be all that clear, but it also happens to be the case that the answer itself is not all that obvious. This is because we know that as numbers get larger, it is in some vague sense "harder" for them to be prime. This is because the larger a number is, the more likely it is for that number to be divisible by some smaller number (simply because the larger a number is, the more numbers there are that are smaller than it). Thus it seems plausible that a number can be so large that it and all of the numbers larger than it can always be divided by smaller numbers. However, when we sit down and start writing down as many prime numbers as we can, it becomes apparent that there are some **very large** prime numbers. For example, it was known as long ago as 1772 that the number 2,147,483,647 is prime. The answer to the question of whether or not there are infinitely many prime numbers was also known in 1772, so let us go on and see for ourselves what the answer is and why it is so.

It turns out there are indeed infinitely many prime numbers, and this has actually been known for millennia. Thus, mathematicians who seek to find larger and larger prime numbers always know that there will be some to find. The mathematician who is credited with the proof that we will give below is Euclid, one of the most influential mathematicians to ever live. The proof uses the method of contradiction. We will **suppose** that there are finitely many prime numbers, so that in particular there is some **largest** prime number. We will then make certain constructions that are totally allowable under the supposition that there are finitely many primes. We will then arrive at the conclusion that there exists a prime number that is larger than the largest prime number. Since this last statement is clearly an impossibility — namely, there is then a prime number which is both the largest and not the largest prime number — it must be the case that our supposition was wrong, and that there are indeed infinitely many prime numbers. Let us see how this goes.

There are two intermediate results that we need to establish before actually proving the main result (that there are infinitely many primes). The first result that we need is that any number in $\mathbb{N} = \{1, 2, 3, ...\}$ is either prime or it is not — no number is both prime and not prime, and no number is neither prime nor not prime. Indeed, if we call numbers that are not prime **composite** numbers, then every number is either prime or composite, not both and not neither. This can be seen simply from the definition of a prime number. In particular, any number is either divisible by something smaller than itself, or it is not. If it is not, then it is prime. If it is, it is composite.

The second intermediate result that we need is that if a number is not prime then it is divisible by a prime number. In the language established above, any composite number is divisible by a prime number. To see this, we note that a composite number is, by definition, divisible by something smaller than itself. If this number that divides our original composite number is prime, then our original composite number is divisible by a prime number and we are done. However, if this number that divides our original number is not prime, then it is **itself** a composite number and is therefore divisible by something smaller than itself. We then continue this process until we finally get a prime number dividing our original number.

EXAMPLE: INFINITELY MANY PRIMES

For example, 25 is composite because it is divisible by 5. Moreover, 5 is prime and therefore we have immediately shown that 25 is divisible by a prime number. Similarly, 44 is composite because it is divisible by 4. And although 4 is not prime, we see that 4 is itself divisible by 2, which **is** prime. Thus, 44 is divisible by 2 and is therefore divisible by a prime number. In this way we can find, for any composite number, some prime number that divides it. We have therefore shown that **any whole number is either prime, or divisible by a prime number**. This will be very important for us in proving the following theorem.

Theorem 13.2. *There are infinitely many prime numbers*

Proof. Let us suppose that there are finitely many prime numbers.[12] If there are finitely many prime numbers, then there must be a largest prime number. Let us call this number[13] P. We note that we do not know the actual value of P — indeed, the whole point of this proof is to show that no such P really exists — all we are doing here is supposing that there is a prime number P, and that every number that is larger than P is **not** prime. If there is such a P, then the following list contains **every** prime number:

$$\{2, 3, 5, 7, 11, 13, 17, 19, ..., P\}.$$

Again we note that we do not know the details of what happens between 19 and P — all that matters is that under our supposition that there is a largest prime number P, the numbers in the above list constitute all of the prime numbers that exist.

Now, since there are finitely many numbers in the above list, we can multiply them all together. There is nothing stopping us from doing this.[14] Let us call this number N, so that N is the result of

---

[12] We will now derive a contradiction.

[13] This is different from our use of the letter P in the previous section. Namely, in the previous section of this chapter P denoted an entire statement. In our current discussion we are letting P denote the hypothetical largest prime number. There is no relation between these two distinct uses of the letter P.

[14] We note that we most certainly **cannot** multiply infinitely many numbers together, because we will almost always get an infinite answer. However, if we multiply any finite amount of numbers together, we will always get some (usually very large, but perfectly reasonable) number as our result.

multiplying 2 times 3 times 5 times 7 times 11 and so on, until finally we multiply it by the number P. In symbols, we have that

$$N = 2 \times 3 \times 5 \times 7 \times 11 \times ... \times P.$$

We note that N may be some immensely large number, but that this does not matter at all — N is just some number that we are allowed to define once we have supposed that there are finitely many prime numbers.

The contradiction now comes when we ask about the number that is one larger than N. Namely, let us examine the number $N + 1$. We can use previously established truths to show that $N + 1$ is a prime number. Namely, we know that every single number has precisely one of the following properties: it is either prime or it is divisible by a prime (not both, not neither). Let us therefore see which of these properties the number $N + 1$ has.

Suppose we tried dividing $N+1$ by any prime number. This means we would have to choose some number from the list $\{2, 3, 5, 7, ..., P\}$ of (supposedly) all prime numbers and see what happens when we divided $N + 1$ by it. Now, we know that N is divisible by **all** of the prime numbers in the list, because N is the product of all of these numbers. Thus, N divided by any prime number would just be the product of all of the numbers in the list $\{2, 3, 5, 7, ..., P\}$ **except** for whichever prime number from the list that we chose. Therefore, the number $N + 1$ will **not** be divisible by any number in the list $\{2, 3, 5, 7, ..., P\}$.

To see this, suppose we chose the number 2 from the list $\{2, 3, 5, 7, ..., P\}$. Then N would be evenly divisible by 2, so that $\frac{N}{2}$ would be a whole number. Namely, since

$$N = 2 \times 3 \times 5 \times 7 \times 11 \times ... \times P,$$

we have that

$$\frac{N}{2} = 3 \times 5 \times 7 \times 11 \times ... \times P,$$

which is a whole number. Therefore, $N + 1$ divided by 2 would be the whole number $\frac{N}{2}$ plus the fraction $\frac{1}{2}$, and thus $N + 1$ divided by 2 is not a whole number and therefore $N + 1$ is not evenly divisible by 2. Similarly, suppose we chose the number 3 from the list

EXAMPLE: INFINITELY MANY PRIMES

$\{2, 3, 5, 7, ..., P\}$. Then N would be evenly divisible by 3, so that $\frac{N}{3}$ would be a whole number. Indeed, since

$$N = 2 \times 3 \times 5 \times 7 \times 11 \times ... \times P,$$

we have that

$$\frac{N}{3} = 2 \times 5 \times 7 \times 11 \times ... \times P,$$

which is a whole number. Therefore $N+1$ divided by 3 would be the whole number $\frac{N}{3}$ plus the fraction $\frac{1}{3}$, and thus $N+1$ divided by 3 is not a whole number and therefore $N+1$ is not evenly divisible by 3. Indeed, suppose we chose **any** number, call it q, from the list $\{2, 3, 5, 7, ..., P\}$. Then N divided by the number q will be a whole number, so that $N+1$ divided by q will be the same whole number plus the fraction $\frac{1}{q}$. Thus $N+1$ divided by q will not be a whole number, and therefore $N+1$ is not divisible by q. And since q could be **any** number in the list $\{2, 3, 5, 7, ..., P\}$, we now know that the number $N+1$ is not divisible by any number in this list.

However, the list $\{2, 3, 5, 7, ..., P\}$ is supposed to contain **all** of the prime numbers. Therefore our conclusion tells us that $N+1$ is not divisible by **any** prime number, and therefore $N+1$ must itself be a prime number (since every number is either divisible by a prime number or it is itself a prime number). However, $N+1$ is larger than P, which is meant to be the largest prime number that exists. By this reasoning, $N+1$ must **not** be a prime number since it is larger than the largest prime number. This means that the statement "$N+1$ is a prime number" is simultaneously true and false.

And that is the contradiction we were after! Since all of this reasoning followed perfectly logically from the **assumption** that there are only finitely many prime numbers, it must be the case that this assumption itself was wrong. Therefore, we now know that there are indeed infinitely many prime numbers, for if there were not, then we would be faced with the absurdity that a certain statement is simultaneously true and false.[15] □

We have therefore established two truths. The first is that there are infinitely many prime numbers, and this is an extremely important truth in the development of mathematics. The second is that

---

[15] That little box in the corner is there to signify that the proof is complete.

the method of proof by contradiction is incredibly powerful. As we will continue to see throughout these volumes, a very wide range of statements can be proved using this method. In the next chapter, we are going to use this method of proof to show that there is a different type of infinity, and in Chapter 15 we will use it to see how there are in fact **infinitely many** different types of infinity. Let us therefore now turn our attention towards establishing these truths, which we have been working towards for several chapters now.

Chapter 14

Infinity Type 2

In Between Fractions

Alas, we have finally gotten to the chapter where we will find a whole new kind of infinity. We have spent many of the last several chapters showing how infinite sets that **seem** either larger or smaller than $\{1, 2, 3, ...\}$ are in fact the same size as it. Of course, by "same size" we really mean "have the same cardinality," and accordingly by "have the same cardinality" we really mean that there exists a bijective function from $\mathbb{N} = \{1, 2, 3, ...\}$ to the set under investigation. In particular, we made the precise definition that a set has "infinity type 1" elements if there exists a bijective function from $\mathbb{N}$ to the set under consideration. We then went on to find that many sets with **seemingly** different cardinalities end up indeed having precisely "infinity type 1" elements. In this chapter however, we will consider a set that is so large that there can never be a bijective function defined from $\{1, 2, 3, ...\}$ to it, no matter how hard we try. Accordingly, we will consider a set with a type of infinity of elements that we have not yet encountered, and it will force us to extend our counting procedure[1] even further.

Before we define this gigantic set, however, we need to first examine the individual elements that we will use to construct it. The

---

[1] Just as we did in Chapter 9 to extend our counting machinery from only being able to handle finite sets to then being able to handle sets with "infinity type 1" elements.

elements in this set will be numbers that go by the technical name "real numbers." Note that a real number is a technical object, and not the opposite of a "fake number." So what exactly is a real number? There are actually many ways that we can rigorously define a real number, and finding these rigorous descriptions of a real number was indeed a hugely important step forward in mathematics.[2] However, the ideas involved in making these rigorous definitions require tools that are far more advanced than those that we have available to us currently. Thus, for now we will simply define a real number in an intuitive and practical manner so that we can **use** the real numbers, even if we do not yet know the deep, abstract, and rigorous definition of them. We can rest assured, however, that such definitions do exist and as we move on in these volumes we will start to be able to unravel those definitions.

For now, we define a real number to be any number that is representable as an infinite decimal. Before quickly reviewing what a decimal representation is all about, we can gain a bit of intuition for real numbers by noting that these are the numbers that make up the "number line" that we may or may not have fond memories of from school. Namely, we may recall constructing "the number line" by drawing a perfectly continuous line[3] with arrows on both ends, which are meant to represent the fact that the line goes off infinitely far in both directions. We place the number 0 on this line somewhere and then view the negative numbers as going off to the left, and the positive numbers as going off to the right, as shown in the following.

A real number is any point at all on this number line. Accordingly, a real number might be a whole number (like 1, 2, 3, or -5), or a

---

[2] And this step forward is due to many remarkable mathematicians in and around the 19th century. The history of these developments are indeed fascinating, but seeing as these volumes intend in no way on covering the history of the ideas presented, we will leave this topic alone for now and simply note that one is encouraged to explore it further.

[3] Here, the line being "continuous" simply means that we never lift our pen or pencil up from the page when we draw it — it is just one solid line.

fraction (like $\frac{1}{2}$ or $-\frac{105}{207}$). However, there are **lots** of numbers on this line (i.e., lots of real numbers) that are **not** fractions. In other words, a fraction is always a real number but the converse is not necessarily true. This has to do with the fact that the numbers on the number line form a genuine **continuum** of numbers, whereas the fractions are in no way "continuous." As mentioned above, we will not be able to make this notion of "continuum" rigorous at the moment, so for now we will rely[4] on our intuitive notion of continuity.

It is indeed good that there are more real numbers than there are fractions, because our aim here is to show that the cardinality of the real numbers is greater than "infinity type 1," which we know from Chapter 12 is the cardinality of the fractions. We will be able to more clearly see the difference between fractions and the non-fraction real numbers once we define real numbers according to their decimal representations. For once we have done this we will then be able to explore the decimal representation of any given fraction and see that there are indeed **tons** of non-fraction real numbers. Let us therefore begin by reminding ourselves of how decimals work.

A number like 1.5 or 12.6 is hopefully not too unfamiliar, but we should refresh our memories regardless. The digits to the left of the decimal point make up the "whole number part" of the real number, and the digits to the right of the decimal point make up the "fractional part" of the number.[5] The whole number and fractional parts of a given decimal can be viewed as giving us instructions on where to find the corresponding number on the number line. To see this, we need to examine the fractional part of a decimal in greater detail.

Once we have used the whole number part of a number to get somewhat close to the corresponding location on the number line, we use each digit to the right of the decimal point in order to approach

---

[4]Let us note that this reliance on intuition is precisely the opposite of what we have worked to establish so far, but for now it is the best we can do. However, this is not so blasphemous because it is almost always the case that one's intuition **guides** them to the more rigorous meaning behind some definition or theorem. Thus, by relying on our intuition for now, we will be in a better position to more fully understand the rigorous descriptions that we will give in a later volume.

[5]In the case of the number 45.2, for example, the number to the left of the decimal is 45 and the number to the right of it is 2.

our number with greater and greater accuracy. For example, the number 2.451 is a symbol that represents some individual point on the number line. This point will be somewhere between 2 and 3, since the whole number part of the number is 2. It is then the fractional part that points us the rest of the way. Namely, we are meant to take the fractional part (in this case it is 451) and form the fraction given by placing the fractional part in the numerator, and a power of ten with as many zeros as there are digits in the fractional part in the denominator. Thus, for this specific example of 2.451, the number in question lies $\frac{451}{1000}^{ths}$ of the way from 2 to 3. Equivalently, we have that

$$2.451 = 2 + \frac{451}{1000}.$$

As further examples, we have that 1.5 is simply "1 and a half," since we have that

$$1.5 = 1 + \frac{5}{10}$$

and that $\frac{5}{10} = \frac{1}{2}$. We can also have negative decimal numbers, in which case we have a **negative** whole number part and then we **subtract** the fractional part. We note that this is in line with what we did with the positive decimals as well. Namely, the whole number part of a positive number gave us very coarse instructions on how far to the right we should walk down the number line. For example, if the whole number part of a decimal is 47 then we know that our number will be between 47 and 48 (i.e., either exactly **on** 47 or somewhere slightly to the right of 47). The fractional part then gave us finer details of where between 47 and 48 our number lies. For negative decimals, the whole number part tells us very coarsely how far to the **left** we should walk, so that if the decimal is -83.64, we know that our number will be somewhere between -83 and -84 (note that this is now to the **left** of -83). We must then continue to walk to the left $\frac{64}{100}^{ths}$ of the way to -84 in order to land on our number. As an example, we have that

$$-12.6 = -12 - \frac{6}{10} = -12 - \frac{3}{5}$$

where in the second equality we used the fact that the fractions $\frac{6}{10}$ and $\frac{3}{5}$ are equal.

This will suffice for our review of decimal notation. So far every decimal that we have written down is **also** a fraction, and so currently it may be unclear as to why we want to introduce this way of writing things. In the next section we will be able to define real numbers in terms of decimal representations, and we will then see that there are lots of possible decimals that are indeed **not** fractions. This is what we will need to be able to see a whole new kind of infinity.

The Working Man's Definition Of Real Numbers

Let us now go on and use our knowledge of decimals to define what a real number is. A real number is, by definition, any number that can be expressed as a decimal, even possibly an **infinitely long** decimal. For example, an expression like

$$145.32639204335123052353123546946274602562645047508509 45...$$

is a real number, where the "..." means that the decimal goes on forever. We note that usually a "..." implies that some kind of well-understood pattern gets repeated forever, the way that the "..." in the definition of $\mathbb{N} = \{1, 2, 3, ...\}$ means that the pattern of increasing whole numbers continues on forever. In the above "..." however, there does not have to be any rhyme or reason to the numbers that are on the right of the decimal place and therefore we might not know which pattern is continuing on forever. This is one of the important aspects of our definition of a real number — the infinitely many digits to the right of the decimal place can continue on forever without ever setting up any kind of pattern.

Let us now make a few remarks about real numbers. Namely, let us first describe how whole numbers and fractions are indeed real numbers.[6] It is rather straightforward to see that any whole number is a real number — a whole number is just a number with zero digits

---

[6] Recall that in our number line discussion, we saw that we want our definition of real number to **include** the whole numbers and fractions. Namely, we know that we want every fraction (and therefore every whole number as well) to be a real number, while it is **not** the case that every real number is a fraction.

to the right of its decimal place. Since 1 and 17 and -45 are all decimals (simply without entries to the right of their decimal places), they are indeed all real numbers as well. We could write 1=1.0 if we wanted to, or even -45=-45.00000 if we wanted to, but there is no need — a real number is any decimal at all (even an infinitely long one), and whole numbers are perfectly good decimals.

In a similar way, it turns out that every fraction can be written as an infinite decimal as well. The rigorous proof of this statement is a little beyond us right now (and would not be particularly fun or enlightening anyway), however the intuitive reasoning behind this statement is the following. It turns out that every fraction either has a decimal expansion that "stops" eventually, or has a repeating pattern eventually. For example, the fraction $\frac{3}{2}$ is nothing but "one and a half," which is expressible as the decimal 1.5. Similarly, the fraction $\frac{3}{8}$ can be represented by the decimal 0.375 because if we multiply 0.375 by 8, we get the number 3, and this is the very definition of the fraction $\frac{3}{8}$. Thus we see that many fractions can be represented as **finitely long** decimals.

However, there are some fractions that can only be expressed as **infinitely long** decimals. For example, $\frac{11}{9}$ is only expressible as the infinitely long decimal 1.222222..., where the 2's keep repeating forever. This is fine, though, because the real numbers include the infinitely long decimals, and so the fraction $\frac{11}{9}$ is indeed a real number. Similarly, the fraction $\frac{1}{7}$ can be expressed as a decimal that has a pattern that repeats over and over again.[7] In particular, it turns out that

$$\frac{1}{7} = 0.142857142857142857...,$$

where the string of numbers "142857" keeps on repeating forever. The method for showing why this is the case is not important to us right now — the important point is the following. **Any fraction at all can be expressed as a (possibly infinitely long) fraction that either "stops" eventually or eventually sets up a pattern that repeats over and over again.** Therefore, **any** infinitely long

---

[7]In the case of $\frac{11}{9}$, the "repeating pattern" was simply the pattern "2." Some fractions, however, may have significantly more complex and interesting patterns that end up repeating forever.

THE WORKING MAN'S DEFINITION OF REAL NUMBERS 131

decimal that never ends and never starts to repeat itself is a real number which is **not** a fraction, and this is why there are **many** non-fraction real numbers. The philosophy of these volumes (and in mathematics in general) is to be as rigorous about every statement as possible. However, we will just have to take the above fact about fractions on faith for now and trust the fact (which is true) that mathematicians have indeed proved it rigorously — we just will not see the proof quite yet.[8]

Before proving that the set of all real numbers is "more infinite" than the previous sets we have seen, let us first remark on why we actually need real numbers. Namely, it is likely that we have no qualms with the existence of whole numbers and fractions, since these kinds of numbers appear regularly in our daily lives. However, we may be somewhat wary of allowing for decimal representations that may be infinitely long and that **never** establish any kind of repeating pattern. These types of numbers may seem somewhat contrived, or artificial. After all, we can never **directly** access all infinitely many decimals of a number whose decimal representation never sets up any kind of pattern — at least not in the way that we can **directly** access the fraction $\frac{2}{3}$ by, say, eating $\frac{2}{3}$ of a pizza. However, there are indeed **tons** of numbers that are very important (and very "real") that simply cannot be expressed as a fraction, and therefore **must** be expressed as an infinitely long — and never repeating — decimal. For example, the numbers "Pi" (which is the famous $\pi = 3.1415926...$, and which is given by taking the total length around a circle and dividing that length by the total length directly across that circle) and the square root of 2 (which is the number that, whenever multiplied to itself, gives 2) **cannot** be written as a fraction.[9] These numbers (as well as infinitely many others) can **only** be written as a decimal that never ends and never repeats, and yet they are extremely important numbers for even the most basic geometric considerations.

Now that we have defined real numbers, seen that they do indeed contain all the fractions, and convinced ourselves that they are in fact necessary, let us now turn our attention towards actually proving that

---

[8]The proof of this is really not all that important for what is to come anyway.

[9]We will provide a proof of the fact that the square root of 2 cannot be written as a fraction in the final chapter of this volume.

there are **wildly** more than "infinity type 1" of them!

Cantor's Diagonal Slash

Let us now consider the set of all real numbers, and let us agree to denote this set by $\mathbb{R}$. Thus, whenever we write $\mathbb{R}$, we mean the set of all numbers (positive or negative) that can be expressed as an infinitely long decimal.[10] In the previous section we saw that the real numbers include the fractions as well as other numbers. However, as we have seen in past chapters, this does not mean that this set has a different cardinality. After all, the fractions include the whole numbers as well as lots of other numbers, yet the set $\{1, 2, 3, ...\}$ and the set $FRAC$ of all fractions have the same cardinality. Let us now establish that indeed $\mathbb{R}$ does have a different — and therefore larger[11] — cardinality than $FRAC$. We will do this by showing that we cannot define a bijective function from $\{1, 2, 3, ...\}$ to $\mathbb{R}$, no matter how clever we are.

But how does one show that something **cannot** happen? Showing that something can happen is easy — we just do it. Showing that something **cannot** happen is more tricky, but luckily we have already developed an extremely powerful tool for proofs of this type. Namely, we can use the method of proof by contradiction. We will assume that we **can** define such a bijective function and then arrive at a logical contradiction, thus showing that this assumption is impossible to make (and thus showing that no such function can exist). The following argument is insanely beautiful and profoundly clever. It is the product of one of the greatest mathematicians to ever walk this planet, who went by the name Georg Cantor.

Let us suppose[12] that there is a bijective function from $\mathbb{N} =$

---

[10] In what is to follow, it is important to remember that a number that can be expressed as a finitely long decimal (like whole numbers and certain fractions) can also be expressed as an infinitely long one by simply attaching infinitely many zeroes after the final digit in the decimal representation. For example, $1.56 = 1.56000...$.

[11] We know that if the cardinality is different, then it must be larger because $\mathbb{R}$ **contains** $FRAC$. We will make this notion more rigorous in the second section of the next chapter.

[12] We will make this supposition, then do things that are totally allowable if

$\{1, 2, 3, ...\}$ to $\mathbb{R}$. Let us call this function F. If such a function exists, we would then have a list of associations between elements in $\mathbb{N}$ and elements in $\mathbb{R}$ such that **every** element in $\mathbb{R}$ is "hit" by an element in $\mathbb{N}$, using the function F. This is because we know F is bijective and therefore surjective.

We are now going to introduce a notation for real numbers that, although somewhat confusing at first, will make the proof extremely easy.[13] We recall that F will assign a real number in $\mathbb{R}$ to each number in $\mathbb{N} = \{1, 2, 3, ...\}$. Thus, in particular, we know that $F(1)$ is some real number. Let us agree to denote this real number by $A_1.a_{11}a_{12}a_{13}a_{14}...$, where "$A_1$" is some whole number (positive, negative, or zero) and it is followed by a decimal point as usual, and each a_{1i} is some number between 0 and 9 (note that the a_{ij}'s are just abstract labels for individual digits). In other words, this is a completely general decimal representation for whichever real number $F(1)$ is, where A_1 is the "whole number part" and each a_{1i} is a single digit of the fractional part. Also, a_{11} should not be thought of as "a" with a subscript of eleven, but rather "a" with a subscript "one-one." Similarly, a_{12} should not be thought of as "a" with a subscript of twelve, but rather as "a" with a subscript of "one-two," and so on.

This notation will become clearer by looking at some examples. If the function F is such that $F(1) = 12.325000000...$, then we would have $A_1 = 12$ since this is the whole number part of $F(1)$, and we would also have $a_{11} = 3$, $a_{12} = 2$, $a_{13} = 5$, $a_{14} = 0$, $a_{15} = 0$, and all the rest of the a_{1i}'s equal 0, for any i that is larger than 5. As another example, let us suppose[14] that the function F is such that $F(1) = \pi = 3.1415926...$. We would then have that $A_1 = 3$, $a_{11} = 1$,

---

this supposition is correct, and then arrive at a contradiction. This will show that this supposition is incorrect. It is therefore important to remember where we make our supposition (i.e., where the **only** place that our argument **might** break down is), so that when our argument **does** break down, we will know why. This footnote is here to remind us that this is the location of our supposition.

[13]We must recall that notation is never scary, it is just new written abbreviations that we need to learn.

[14]Remember, we do not know what the details of F are — namely, we do not know where F **actually** sends the elements of $\mathbb{N}$. All we have done so far is suppose that there **exists** a function F from $\mathbb{N}$ to $\mathbb{R}$ that is bijective. Thus, we need to develop a notation that accounts for **any** possible such F, and this is what we are doing now.

$a_{12} = 4$, $a_{13} = 1$, $a_{14} = 5$, $a_{15} = 9$, $a_{16} = 2$, $a_{17} = 6$, and so on, according to the infinitely long (and never repeating) decimal representation of π. We therefore see that for **whatever** real number $F(1)$ might be, we will be able to represent it by $A_1.a_{11}a_{12}a_{13}a_{14}...$, where A_1 is a whole number (positive, negative, or zero) while each a_{1i} is some digit between 0 and 9.

Let us therefore agree to similarly denote the real number $F(2)$ by $F(2) = A_2.a_{21}a_{22}a_{23}a_{24}...$ where now A_2 is some whole number (positive, negative, or zero) and each a_{2i} is some digit between 0 and 9. In fact, we can quickly see how to generalize this notation to represent the real number $F(i)$ for whichever number i in $\mathbb{N}$ we may be considering. All we need to do is use the number i, which is an element in $\mathbb{N}$, to denote the whole number part A_i of $F(i)$. Then, we use this same number i as the number in the first subscript of each "a_{ij}," so that i is simply a bookmark for which element in the set $\mathbb{N}$ that we are talking about. Finally, the second subscript of each "a_{ij}" (namely, j in this case) is simply a bookmark for how far to the right of the decimal point we are. Thus, for each number i in the set $\mathbb{N} = \{1, 2, 3, ...\}$, and for whatever our function F does to these numbers,[15] we can write $F(i) = A_i.a_{i1}a_{i2}a_{i3}a_{i4}...$. Then, A_i is simply the whole number part of the real number $F(i)$, and each a_{ij} is simply the digit in the jth place of the decimal representation of the real number $F(i)$.

This notation may be confusing upon seeing it for the first time, so it may take a second to sink in. However, once it sinks in, we will have a mathematical sword that is sharp enough to take this proof down in one thrust.

We recall that so far all we have done is **suppose** that there is a function $F : \mathbb{N} \to \mathbb{R}$ that is bijective, and then introduced a useful (albeit possibly confusing) notation to represent this function, whatever it may be. What we will do now is find a real number that cannot possibly be "hit" by this function F. This is impossible though, because the function F was **supposed** to be bijective, and therefore surjective, and it therefore should hit every single real number! This is the contradiction that we are after, so let us now go on to find this

---

[15] Namely, wherever these numbers get sent to in $\mathbb{R}$.

number that is not be hit by F.

Given such a hypothetical function F, we can construct the infinitely long list that is shown in Figure 14.1. We recall that for any actual (i.e., explicit) F, this will be a list of explicit real numbers. Here, we are simply allowing for F to be **any** (supposedly bijective) function from $\mathbb{N}$ to $\mathbb{R}$. We will now use this list to find a real number that cannot possibly be in the list — no matter what F is — and we do this using Cantor's "diagonal slash" argument. Once we find a real number that cannot possibly be in the list, then we will have found a real number that our function F has missed, since this (infinitely long) list contains all of the real numbers that F hits.

$$F(1) = A_1.a_{11}a_{12}a_{13}a_{14}a_{15}a_{16}...$$
$$F(2) = A_2.a_{21}a_{22}a_{23}a_{24}a_{25}a_{26}...$$
$$F(3) = A_3.a_{31}a_{32}a_{33}a_{34}a_{35}a_{36}...$$
$$F(4) = A_4.a_{41}a_{42}a_{43}a_{44}a_{45}a_{46}...$$
$$F(5) = A_5.a_{51}a_{52}a_{53}a_{54}a_{55}a_{56}...$$
$$\vdots \qquad \vdots$$

Figure 14.1: A list that goes infinitely far down and infinitely far off to the right, enumerating the associations that the function $F : \mathbb{N} \to \mathbb{R}$ gives.

We begin by noting that any real number can be written as

$$B.b_1b_2b_3b_4b_5...,$$

where B is a whole number (positive, negative, or zero), and each b_i is a digit in the set $\{0, 1, 2, ..., 9\}$. This is exactly the same as what we have done above with the numbers $A_i.a_{i1}a_{i2}a_{i3}...$, only without the extra subscript on the b_i's because now we only need to represent one real number (as opposed to the infinitely many real numbers that our function F hits). Let us therefore consider the following real number. Let us consider the real number with $B = 0$, (i.e., a real number with only a fractional part), and let us define b_1 to be $a_{11} + 1$. In other words, we take whatever numerical value a_{11} has in the list above[16] and add 1 to it. For completeness, since each a_{ij} must be between 0 and 9 (and therefore we cannot really add 1 to 9 without going out of this range), let us say that if a_{11} equals 9, then we make b_1 equal 0. Similarly, let us define b_2 to be $a_{22} + 1$, (and if $a_{22} = 9$ then we make $b_2 = 0$), b_3 to be $a_{33} + 1$, b_4 to be $a_{44} + 1$, and in general let us define b_i to be $a_{ii} + 1$, where for all of these b_i we agree that if $a_{ii} = 9$ then we make $b_i = 0$. Note that all we are doing is going down the above list **along the diagonal**[17] and changing each digit on this diagonal by adding 1 (or changing it to 0 if it started out as 9). In this way, we have defined a perfectly good real number — namely, the number $0.b_1 b_2 b_3 b_4...$ — and we have done so for **any** explicit values that our function F takes (namely, for whatever explicit real numbers show up on the right hand side of the above list). Let us call this real number P, so that $P = 0.b_1 b_2 b_3 b_4....$

Let us now ask where P is in our list above. It must be **somewhere** in our list, because P is a real number and our list contains **all** real numbers (by assumption of the fact that F is bijective and therefore surjective). But P cannot be the first number in that list because it differs from $F(1)$ in the first decimal place — b_1 does not equal a_{11} because $b_1 = a_{11} + 1$! P also cannot be the second number in our list because it differs from $F(2)$ in the second decimal place (since $b_2 \neq a_{22}$, by definition of b_2), and it cannot be the third number in our list because it differs from $F(3)$ in the third decimal place (because $b_3 \neq a_{33}$, by definition of b_3). Indeed, suppose our number P were **anywhere** in the above list. It would then have some position in our list, so let us call this position N. This would mean that $F(N)$ equals P, which is impossible because b_N is the Nth decimal place of P and it differs from a_{NN}, which is the Nth decimal place of $F(N)$. Thus it is simply impossible for P to be in our list — it is a real number that is defined to differ from **every** real number in our list in at least one decimal place.[18]

---

[16] We note that we do not know what this numerical value is, since we do not know the explicit form of F. The point is, though, that **for any** F that we could possibly have, we will always be able to do this. In other words, for **whichever** value a_{11} has, we can always consider the number b_1 which is 1 more than it.

[17] This is why it is called Cantor's "diagonal slash" argument.

[18] A question one may ask is why we cannot simply take this real number P

Thus our function, which supposedly hits every real number, missed at least one real number! This is clearly a contradiction, for a function cannot simultaneously hit every real number as well as **not** hit every real number!

Therefore such a bijective function F cannot exist, because solely by supposing that it did we were able to come to a conclusion that is completely logically impossible. Thus, since there absolutely cannot ever be a bijective function from $\mathbb{N} = \{1, 2, 3, ...\}$ to the set $\mathbb{R}$ of all real numbers, it must be the case that $\mathbb{R}$ has a different cardinality than $\mathbb{N}$. However, since $\mathbb{R}$ is most definitely an infinite set, and since $\mathbb{R}$ **contains** the set $\mathbb{N}$ as a subset, it must be the case that the cardinality of $\mathbb{R}$ is **larger**[19] than "infinity type 1." We have therefore found our new infinity, and we see that indeed it must be larger than "infinity type 1"!

What we have found, therefore, is that even though we have extended our number system by including "infinity type 1," we still have sets that we cannot count. Namely, any set that can have a bijective function defined from $\mathbb{N} = \{1, 2, 3, ...\}$ to it is called **countable**, because we can label every element in that set with a number. Accordingly, we call a set that is infinite but that **cannot** have a bijective function defined from $\{1, 2, 3, ...\}$ to it an **uncountable** set. In the next chapter, we will see that there is even more to the story. Namely, it is not even the case that there are only two different kinds of infinities. Instead, we will see that there is an infinitely tall tower of infinities, each larger than the ones before it.

For now, let us reflect on how far we have come, and celebrate the remarkable fact that we have already been able to use our abstract

---

and "add it in" to the right hand side of the above list. Namely, why do we not just consider a new function that hits P first and simply "shifts" all of the other assignments up one, as we did with our infinite hotel when we wanted to add in one friend. For if we can do this then our function will have indeed "hit" P and it would seem like our problems go away. However, this would then be a **new** function and we would therefore be able to use the same logic that we used here to construct **yet another** real number which this new function does not hit (by using the exact same diagonal slash argument). Thus, we will **always** be missing at least one real number, no matter how hard we try.

[19] We will have more to say about "which infinity is larger" in the second section of the next chapter.

mathematical machinery to prove that there is, in a very well-defined and definitive way, more than one type of infinity. It took some work, but that work led us to a truth that is profound, surprising, and completely eternal. This is the true beauty of math — to employ our mental facilities along with a pristine logical framework to gaze upon truths that exist in an abstract and eternal world that is not at all bound to the physical, sociological, or economic constraints of any civilized being. Let us therefore, after a moment of reflection, continue on into this wonderful world and gaze upon ever more beautiful truths.

Chapter 15

Infinity Infinities

Contrapositives

In the previous chapter we established that there are at least two different kinds of infinities, and we will now go on to see that there are in fact infinitely many different kinds of infinities. The method of proof that we used in the previous chapter is very similar to that which we will use in this chapter, so that fully understanding the jump from one to two infinities (which we covered in the previous chapter) will make understanding the jump from two to infinity infinities a breeze. In particular, we will again use the method of proof by contradiction, and we will also use an idea very similar to the "diagonal slash" that Cantor introduced and which we used in the previous chapter.

Recall that this "diagonal slash" method involved **supposing** that there is a bijective function from the infinite set $\mathbb{N}$ to the infinite set $\mathbb{R}$, and then finding an element in $\mathbb{R}$ that is not "hit" by the function. We found this element by going down the "diagonal" and defining a real number in a way that ensures that it does not show up anywhere in the list that represents the function. Then, since this is a contradiction to the fact that this function should have been bijective (and therefore surjective), we knew that our supposition must have been wrong and indeed that there **cannot** be a bijective function from $\mathbb{N}$ to $\mathbb{R}$. We were then forced — by definition of what it means to have cardinality "infinity type 1" — to acknowledge the fact that

$\mathbb{R}$ does not have "infinity type 1" elements.

There are two questions that remain, however. First, we showed that there is no bijective function from $\mathbb{N}$ to $\mathbb{R}$, but does that mean that there is no bijective function going the other way, namely from $\mathbb{R}$ to $\mathbb{N}$? Surely if there were such a function we would still want to say that $\mathbb{R}$ has the same cardinality as $\mathbb{N}$, so can we really be so quick to say that their cardinalities are different? Second, even if there is not such a bijective function going the other way, how can we be sure that the infinity of the number of elements in $\mathbb{R}$ is **larger** than the infinity of the number of elements in $\mathbb{N}$? Both of these questions are genuine, in the sense that they truly need be dealt with before we can continue. Fortunately, they both have straightforward answers. We will unravel the answer to the first question in the remainder of this section, and then we will discuss the answer to the second question in the next section. These questions (and their answers) will then be extremely important for showing that there are indeed infinitely many infinities, each larger than the one before it.

To begin, let us consider the question of whether or not we can define a bijective function "the other way" when we could not do so the first way. Namely, suppose we find ourselves in the situation where we have two sets A and B, and we know that there cannot ever be a function $f : A \to B$ that is bijective.[1] We want to know whether or not it is possible to define a bijective function $g : B \to A$. It turns out that the answer is no — if we cannot define a bijective function one way, then we also cannot do so the other way. The way that we will prove this is by proving the **contrapositive** statement. Namely, we will prove that if we **can** define a bijective function one way, then we also **can** define one the other way. This is completely equivalent to proving that if we **cannot** do so one way, then we also **cannot** do so the other way. Let us take a few paragraphs to introduce and discuss this "contrapositive logic," as it is extremely important.

Suppose we have two statements P and Q. Here, P is the statement "there is no bijective function from the set A to the set B" and Q is the statement "there is no bijective function from the set B to the set A." Suppose we want to prove that the statement P implies

---

[1] Note, this is exactly the situation in which we find ourselves, where now $A = \mathbb{N}$ and $B = \mathbb{R}$.

the statement Q. Namely, suppose we want to prove that "if P is true, then Q must be true." Let us agree to denote this by $P \Rightarrow Q$, so that the fancy arrow "$\Rightarrow$" simply reads "implies" and the symbols "$P \Rightarrow Q$" reads as "P is true implies Q is true." Note that in our case, we want to prove that "there is no bijective function from A to B" implies "there is no bijective function from B to A."

Now for any two statements P and Q, there are two ways to prove that $P \Rightarrow Q$. Namely, there are two ways to prove that "if P is true then Q is true" (assuming that P actually does imply Q, of course). We can either prove it directly by using some set of logical deductions beginning with the truth of P and arriving at the truth of Q, or we could prove the **contrapositive** of the statement. The contrapositive of the statement $P \Rightarrow Q$ is, by definition, the statement "not $Q \Rightarrow$ not P." Here, "not Q" simply means "the opposite of Q," (so that if "not Q" is true then Q is false). For example in our case, where P is the statement "there is no bijective function from the set A to the B," we have that "not P" is the statement "there is a bijective function from the set A to the set B." Similarly, since Q is the statement "there is no bijective function from the set B to the set A," we have that "not Q" is the statement "there is a bijective function from B to A."

To recap, the contrapositive of the statement "if P is true then Q is true" is the statement[2] "if Q is false then P is false." What we are going to prove shortly, and which we will use extensively, is that if we can prove the contrapositive of a given statement, then we have equally well proven the statement itself. Namely, if we can prove the statement "if Q is false then P is false," then we will have equally well proven that "if P is true then Q is true." Before showing **why** these two statements are equivalent, let us first introduce a bit more notation. Let us denote the statement "not Q" by "$\neg Q$" just to save us a bit of writing (and to make everything look a bit cooler). Thus, the two expressions "not $Q \Rightarrow$ not P" and "$\neg Q \Rightarrow \neg P$" are simply two different ways of denoting the exact same statement.

Let us now see why proving that "$P \Rightarrow Q$" is completely the same as proving that "$\neg Q \Rightarrow \neg P$." For suppose that we were able to prove that "$\neg Q \Rightarrow \neg P$." This means that we have **already been able to**

---

[2]The switching of the order of P and Q when we consider the contrapositive is not an accident — it is extremely important for this logic to work.

prove that whenever Q is false, P is false as well. Then, if we are ever in a situation where P is true, we will automatically know that Q is true too. This is because if Q were false (when P is true), then we could use the **already proven** statement that "if Q is false then P is false" to prove that P is false. But this is a contradiction to the fact that we know we are in a situation in which P is true! Thus, if we can prove that "if Q is false then P is false also," then we know that "if P is true then Q must be true also." The latter of these two statements is precisely the statement "$P \Rightarrow Q$," and the former is precisely the statement "$\neg Q \Rightarrow \neg P$." Therefore, if we can prove one of these two statements (either the statement itself or the contrapositive), then the other one follows immediately.

In our case, we want to prove that if there is not a bijective function from A to B, then there is not a bijective function from B to A. What we have just seen is that we can equally well prove that if there **is** a bijective function from B to A, then there **is** a bijective function from A to B. For if we can prove the latter statement, then we will have proved the former statement as well (because the latter is the contrapositive of the former). Let us therefore go ahead and prove the following proposition.[3]

Proposition 15.1. Let A and B be sets. If there exists a bijective function $f : A \to B$, then there also exists a bijective function $g : B \to A$.

The idea of the proof will be the following. If we have a bijective function $f : A \to B$, then that means that we could in principle draw an arrow from each element of A to some element of B in such a way

---

[3]Some of our results will be labeled as "propositions" and some of them will be labeled as "theorems." There is very little distinction between the two — both are simply mathematical statements that need proving. Mathematicians also sometimes use the word "lemma" to label a mathematical statement, and all of these words have very similar meanings. Theorems are generally more substantial statements either in regards to their importance or in regards to the difficulty of their proof. That said, there are "lemmas" that great mathematicians have worked their entire lives to prove, and "theorems" that are rather trivial. We will therefore be rather careless in whether we label a certain mathematical fact by "Theorem," "Proposition," or "Lemma." All three of these terms simply mean that the corresponding statement is some mathematical truth that either has already been proved or is about to be proved.

that no element of B is hit twice, and such that no element of B is missed. Note that Figure 6.3 gives an example of this. In general, if there is an arrow from an element a in A to an element b in B, then that means that $f(a) = b$, i.e., that our function f assigns the element b in B to the element a in A. We then see that if we simply reverse the direction of each arrow, we will have a perfectly good bijective function from B to A. Namely, we will have a set of arrows from **every** element of B to the elements of A in such a way that no element of A is hit twice, and no element of A is missed.

It is important to note that we can reverse the direction of the arrows of **any** function, but that we will not always get a function from B to A as the result of doing this. Namely, if we look at Figure 6.1, we see that if we reverse the direction of the arrows we **do not** get a function as the result, because this "function" from B to A would not send c (in B) anywhere in A. Recalling that a function must send **every** element of its domain to **some** element of its codomain, we see that reversing the arrows of an injective function does not result in a function unless the original function was also surjective (and therefore bijective). Similarly, if we look at Figure 5.1, which represents a surjective function, we see again that reversing the direction of the arrows **does not** give a function from B to A. This is because upon reversing the arrows we see that the element b in B is sent to **both** 1 and 2 in A. Since a function must send every element in its domain to **only one** element in its codomain, the "function" from B to A defined by reversing the arrows in this figure is not really a function at all. Thus, reversing the arrows of a surjective function does not result in a function at all unless the original function is also injective (and therefore bijective).

In order to formally prove the above proposition, we need to see what "reversing the arrows" means mathematically. Suppose that $f : A \to B$ is a bijective function from A to B. As mentioned above, there is an arrow from a in A to b in B if $f(a) = b$. What we want is a function $g : B \to A$ that corresponds to reversing the arrows, which means that it must send b in B to a in A. Thus, we need to define $g : B \to A$ by the assignments $g(b) = a$ if it was the case that $f(a) = b$. Remember, f is the **given** function, and we are simply using it to construct $g : B \to A$ according to the following rule. If

b in B is such that a in A is mapped to it by f, then send b to a using g. We can then use the fact that f is bijective to prove that g is bijective, and we will do so in the following formal proof of the above proposition.

Proof. Let $f : A \to B$ be a bijective function, which exists due to the assumption of the proposition. Let us now define the function $g : B \to A$ by $g(b) = a$ if it is the case that $f(a) = b$. This is well-defined as a function, in the sense that it does send every element of B somewhere in A due to the surjectivity of f, and it also sends each element of B to only one element of A due to the injectivity of f.

Let us now see that g is indeed bijective. We will first show that it is surjective. Namely, suppose a is some element (any element) in A. We want to show that there is some element in B that hits this element using the function g. However, since a is in A, it must be sent somewhere in B by the function f, so let us call this element b. In other words, b in B is such that $f(a) = b$. By definition of g, then, we know that $g(b) = a$. Thus there is some element in B that gets sent to a in A by g. Since a in A is an arbitrary element, we see that g is surjective.

Finally, let us show that g is injective. Namely, we need to show that if $g(a) = g(b)$ where now a and b are both[4] elements in B, then they must actually be the same element, i.e. $a = b$. Recall that this is how we phrase the fact that g does not send two different elements in B to the same element in A. So let us suppose that there are elements a and b in the set B such that $g(a) = g(b)$. Let us denote this element in A — i.e., the element in A to which g sends both a and b from B — by c. We then have that $g(a) = g(b) = c$. However, by definition of g, if c is the element in A that g sends a to, then a must be the element in B that f sends c to. In other words, it must be the case that $f(c) = a$. Similarly, since $g(b) = c$ as well, we must have that $f(c) = b$. Thus we have found that $a = f(c)$ and that $b = f(c)$, and therefore it must be the case that $a = b$. We have therefore shown that if $g(a) = g(b)$, then it must also be the case

---

[4]This is important! In the previous paragraph a was some arbitrary element in the set A, whereas now it is simply denoting some arbitrary element in the set B.

that $a = b$, and this is precisely the definition of injectivity! We have therefore shown that g is not only a well-defined function, it is also bijective. This completes the proof. □

We note that the actual truth of the proposition might have been clear from our discussion preceding this proof. Namely, our discussion of "reversing the arrows" of a bijective function may have been sufficient for convincing ourselves that the proposition was true — that if we reverse the arrows we get a perfectly good bijective function "the other way." However, it is always important to make our proofs completely water-tight, and this is what we have done (even if it seems rather unnecessary). It is important to take some time to reflect on not only **how** we presented the various statements in this proof, but also **why** we had to do so.

By formally proving Proposition 15.1, we have obtained an even more general result than we were after. We wanted to show that since there is no bijective function from $\mathbb{N}$ to $\mathbb{R}$, then there must be no bijective function from $\mathbb{R}$ to $\mathbb{N}$. Instead what we have shown is that if there are **any two sets** A and B such that there is no bijective function from A to B, then there is also no bijective function from B to A. We did so by proving the contrapositive, because if there **is** a bijective function one way then we **know** that there is a bijective function the other way. Thus, once we know that there **is not** a bijective function one way, then we also know that there **is not** a bijective function the other way. Therefore, we now know that there is **not** a bijective function from $\mathbb{R}$ to $\mathbb{N}$, and therefore that $\mathbb{N}$ and $\mathbb{R}$ really do have different cardinalities! We now need to go on to the next section to see how it is that we really know which of their cardinalities is bigger.

Let us finish this section by quickly introducing a new bit of terminology. Namely, now that we know that if there is a bijective function one way between two sets then there is also a bijective function the other way, we can summarize this "two-way-ness" of bijectivity by simply saying that the two sets can be put in "bijective correspondence" with each other. Namely, saying that there is a bijective function from a set A to a set B is a very "one-sided" statement, because it emphasizes where we start and where we finish. However, we know that we can "turn this function around" if we want to, and obtain a bijective function from B to A. It is important to remember this fact about bijectivity — that it can be a two-way street if we want it to — and so we can summarize all of this by saying things like "there can be a bijection **between** A and B," as this does not necessarily specify any directionality. Of course, once we want to consider a **particular** bijective function, then we **do** need to specify a direction — i.e., a domain and a codomain — because any **single** function must indeed go only one way.

Which Infinity Is Bigger?

Let us return to the specific case of considering the two sets $\mathbb{N}$ and $\mathbb{R}$. We saw in the previous chapter that we cannot define a bijective function from $\mathbb{N}$ to $\mathbb{R}$, and we saw in the previous section that this means we also cannot define a bijective function from $\mathbb{R}$ to $\mathbb{N}$. This means that we are indeed correct in saying that these two sets have different cardinalities, however we are still faced with a problem. Namely, both $\mathbb{N}$ and $\mathbb{R}$ are clearly infinite sets, so their cardinalities must both be some kind of infinity. We know that, by definition, the cardinality of $\mathbb{N}$ is "infinity type 1," and so we know that the cardinality of $\mathbb{R}$ is **not** "infinity type 1." Precisely determining the cardinality of $\mathbb{R}$ is a fascinating question that we will pay more attention to later in this chapter, but for now let us simply denote this cardinality by $\mathcal{C}$. Thus, $\mathcal{C}$ is just some infinite number which we know is not equal to "infinity type 1." The problem that we are faced with, then, is determining in a consistent fashion which cardinality is larger. We note that this may be clear, in the sense that it just **seems** like there are **more** real numbers than there are whole numbers (especially because $\mathbb{N}$ is contained in $\mathbb{R}$), but we need to make precise **why** we feel this way.

As usual, let us gain some intuition from the case of finite sets. Namely, for concreteness, let us consider the "counting sets" of the form $\{1, 2, 3, ..., N\}$ for some N. These are the sets that we used in Chapter 8 when we defined our new abstract way of counting. Suppose we are given two such sets. Namely, suppose we have a set $\{1, 2, 3, ..., N\}$ and another set $\{1, 2, 3, ..., M\}$, where M and N may or may not be equal. By definition of our counting, we know that $M = N$ **if and only if** there is a bijective function from one of these sets to the other.

However, suppose N is not equal to M, so that these two sets cannot be put into bijection with each other. How do we know which set has a larger cardinality? Well, for finite numbers like N and M we simply need to ask which of these two numbers is larger. But if we look more closely at what we **really mean** when we ask which of these two numbers is larger, we will find the right definition of "larger number" that we can use to generalize to the case of infinite numbers.

In particular, let us suppose for the moment that $N = 5$ and $M = 4$, so that the question we are asking here is which set is the larger between $\{1, 2, 3, 4, 5\}$ and $\{1, 2, 3, 4\}$. **Obviously** it is the case that $\{1, 2, 3, 4, 5\}$ is larger because 5 is greater than 4. But suppose we did not know this.[5] If we did not know that 5 was larger than 4, then we could still say that $\{1, 2, 3, 4, 5\}$ has the larger cardinality by noticing that there **does** exist a **surjective** function from $\{1, 2, 3, 4, 5\}$ to $\{1, 2, 3, 4\}$, but not vice versa. Namely, $\{1, 2, 3, 4, 5\}$ can completely "cover" the set $\{1, 2, 3, 4\}$, whereas $\{1, 2, 3, 4\}$ cannot "cover" $\{1, 2, 3, 4, 5\}$.

If we now consider the general case of $\{1, 2, 3, ..., N\}$ and $\{1, 2, 3, ..., M\}$ again, then we can quickly generalize the discussion of the preceding paragraph for determining which set is larger (if they are not equal). Namely, if we know that $N \neq M$ (i.e., if we know that there is no bijection between these two sets), then we say that $\{1, 2, 3, ..., N\}$ has the larger cardinality if there exists a **surjective** function from $\{1, 2, 3, ..., N\}$ to $\{1, 2, 3, ..., M\}$. Otherwise, we say that $\{1, 2, 3, ..., M\}$ has the larger cardinality. We note that this is perfectly in line with our intuition, because if two sets cannot be put into bijection with each other then it must be the case that one of them "covers" the other — i.e., it must be the case that there is a surjective function **from** one of them and **to** the other (but not vice versa). The set **from which** there is a surjective function is then

---

[5]We want to suppose we do not know this because then we align ourselves with the problem that we are facing with infinite cardinalities. Namely, we know that "infinity type 1" is not equal to $\mathcal{C}$, but we do not immediately know which of these two is larger.

defined to be the larger set.

We are now in a position to rigorously define which cardinality is larger between "infinity type 1" and $\mathcal{C}$. Namely, we can certainly define a **surjective** function from $\mathbb{R}$ to $\mathbb{N}$. For example, since each whole number is also a real number, we can simply send 1 in $\mathbb{R}$ to 1 in $\mathbb{N}$, and 2 in $\mathbb{R}$ to 2 in $\mathbb{N}$, and 3 in $\mathbb{R}$ to 3 in $\mathbb{N}$, and so on. Then, we simply send all the other real numbers to anything we would like in $\mathbb{N}$, since this function is already surjective. Thus, since there is a surjective function from $\mathbb{R}$ to $\mathbb{N}$, and since there is no bijection between them, we can safely say that $\mathcal{C}$, the cardinality of $\mathbb{R}$, is **larger** than "infinity type 1."

We note again that even though this result may seem obvious (or perhaps not), it is important to have it formalized before moving on to the next section when things get even crazier. Namely, we have now made clear what we mean when we say that one infinity is larger than another. In particular, two infinite sets have different cardinalities when they cannot be put into bijection with each other. In this case, the **larger** of the two sets is the set that can have a **surjective** function **from** itself to the other infinite set. With this framework in our toolbox, we will now turn our attention towards uncovering an infinitely high tower of ever increasing infinities, where each infinity cannot be put into bijection with any of the infinities "below it," and where each infinity "covers" the infinities below it.

The Next Infinity

The way that we will show that there are infinitely many different infinities is very similar to the way that one shows that there are infinitely many numbers. We know that there are infinitely many numbers because if the enemy hands us some number, we know that we can always add 1 to it to get a new number that is different from all the numbers below the number that the enemy handed us.[6] Then, since "some number plus 1" is again a number, we can add 1 to **this** number as well and find yet another new number that is different

---

[6]In fact, we can add a million to that number if we want to, but adding 1 is sufficient.

THE NEXT INFINITY 149

from all the numbers that come before it. Since this process never has to end — namely, since there is no finite number that we **cannot** add 1 to in order to get a finite number that is different from (and bigger than) all the numbers that come before it — we know that there are infinitely many finite numbers.

In a similar way, we will now show that if the enemy hands us an infinite set, we can define a **new** infinite set whose cardinality is fundamentally bigger than all of the infinite sets that came before it.[7] We will be able to do this by finding the analog of "adding 1" to the infinite cardinalities — namely, by taking any infinite set at all and **creating a new** infinite set with a fundamentally larger cardinality. Then, since this new set will also be infinite, we will be able to perform this operation again, namely create **yet another** new, and fundamentally larger infinite set. This will land us on yet another infinite set and so we can repeat the procedure yet again. Since this process of taking an infinite set and creating a new infinite set with a larger cardinality never has to stop, we will see that there are indeed infinitely many larger and larger infinite cardinalities, just as there are infinitely many larger and larger finite cardinalities. The starting point for all of this, though, is finding this method of "adding 1" to infinite sets, so let us see how this works.

We begin with the first infinite cardinal that we encountered, which we have been calling "infinity type 1." It is important for us to understand that this is the "smallest" infinite cardinal, in the sense that **any** infinite set can have a **surjective** function defined from itself to $\mathbb{N}$. The reason for this is the following. Suppose we are given **any** infinite set, and suppose we call it A. Since A is infinite, there must be at least a **subset** of A that we can label as $\{a_1, a_2, a_3, ...\}$, were each a_i is some different element of A. For example, if A was the set $\mathbb{R}$ of real numbers, these elements might be the numbers $\{1, 2, 3, ...\}$, or possibly something more chaotic like $\{.4, \pi, 5.6, -9, ...\}$. The important point is that there **must** be some subset of A that we can label in this way, if A is truly infinite. Then, we can simply define the surjective function that sends a_1 (whatever

---

[7]And any infinite set has a larger cardinality than any finite set, so that this new set will actually have a larger cardinality than **all** of the sets that came before it (finite or infinite).

this element might be) to 1, and a_2 to 2, and a_3 to 3, and so on. This is a surjective function, and therefore A must have a cardinality of **at least** "infinity type 1." If there also happens to be a **bijective** function from A to $\mathbb{N}$ (or vice versa), then we know that A has cardinality of **exactly** "infinity type 1," but in general we know that A always has at least "infinity type 1" elements, so long as A is infinite.

Let us therefore see if we can start with the cardinal "infinity type 1" and try to figure out how to "add 1" to it. It turns out that what we need for this task is the notion of a **power set**, which we learned about in Chapter 3. Let us quickly recall this construction here.

Recall that if we have a set A (any set at all, finite or infinite), we can form its power set, which we denote by $P(A)$. The power set $P(A)$ of a set A is the set whose elements are all of the **subsets** of A. We saw in Chapter 3 that if A happens to be a **finite** set with N elements, then $P(A)$ is also a finite set with 2^N elements.[8]

However, regardless of whether or not the set A has finitely many elements, we can **always** define its power set. For example, if $A = \mathbb{N} = \{1, 2, 3, ...\}$, there is no reason why we **cannot** define the set of all subsets of A. Clearly, if A has infinitely many elements, so does its power set $P(A)$. This is easily seen by realizing that each element in A (of which there are infinitely many) defines a perfectly good 1-element subset of A, and therefore defines a perfectly good element in $P(A)$. Taking $A = \mathbb{N}$ as our example again, the previous sentence simply means that since $\{1\}$ is a perfectly good subset of A, as is $\{2\}$, $\{3\}$, $\{4\}$, and so on, we see that we already have infinitely many 1-element subsets of A — all of which are **elements** in $P(A)$. Thus, we see that $P(A)$ is infinite whenever A is infinite. However, it is important to notice that there are **way more** subsets that make up the elements of $P(A)$ than just the 1-element subsets! This is because we also have all of the possible 2-element subsets of A as elements of $P(A)$, as well as the possible 3-, 4-, and 5-element subsets, and indeed also the N-element subsets for any finite number N. It gets even better though, because A also has infinite-element subsets[9] that

---

[8]See Chapter 3 for a description of the notation 2^N.

[9]For example, if $A = \mathbb{N}$, the odd numbers are an infinite subset, as are the even numbers, as are the primes, as are the multiples of 3, the multiples of 4, the multiples of 5, and so on. We could also take away any finite number of elements

make up some of the elements of $P(A)$ as well!

Let us continue to focus on the case $A = \mathbb{N} = \{1, 2, 3, ...\}$ and ask whether or not A can be put into bijection with $P(A)$. In particular, as we saw in the previous paragraph, there are "many more" elements in $P(A)$ than there are in A, in the naive sense of seeing that every element in A forms a 1-element subset of $P(A)$, and that there are many **other** elements in $P(A)$ as well. Thus it **seems** like the cardinality of $P(A)$ is larger. However, we must recall that every whole number was also an element of $FRAC$, and that $FRAC$ had many other non-whole number elements, and yet we showed in Chapter 12 that the cardinality of $FRAC$ and $\mathbb{N}$ were actually the same. Therefore we need to be thorough and actually investigate whether or not there is a bijective function from A to $P(A)$ when $A = \mathbb{N}$, and not simply rely on our guesses. Indeed, we will find that the answer is no — the cardinality of $P(A)$ is larger than that of A — so let us see why.

Our weapon of choice for showing that there is **not** a bijective function from[10] $\mathbb{N}$ to $P(\mathbb{N})$ will be proof by contradiction. Let us therefore **suppose** that there **is** a bijective function, call it F, from $\mathbb{N}$ and $P(\mathbb{N})$. If we could ever have such an F, then this would mean that F maps each element (i.e., positive whole number) in $\mathbb{N}$ to some **subset** of the positive whole numbers, since these are what make up the elements of $P(\mathbb{N})$. We would therefore have infinitely many expressions of the form $F(1) = S_1, F(2) = S_2, F(3) = S_3, ...$, where S_1 is an **entire subset** of $\mathbb{N}$, as are S_2, S_3, and so on. Namely, S_1 is the **subset** of $\mathbb{N}$ (or equivalently, the element in $P(\mathbb{N})$) that our hypothetical function F sends 1 to, and in general we simply denote by S_i the subset of $\mathbb{N}$ that our function F sends the element i in $\mathbb{N}$ to.

For example, let us suppose our function F sends the number 1 in $\mathbb{N}$ to the subset $\{4, 12, 80\}$ in $P(\mathbb{N})$, the number 2 in $\mathbb{N}$ to the

---

in A and take what is left to form a new infinite subset of A. Indeed, we see that there are infinitely many infinite subsets of any infinite set.

[10] For the remainder of this example, since $A = \mathbb{N}$, we will write $\mathbb{N}$ and $P(\mathbb{N})$ instead of A and $P(A)$ simply to remind us that this is a **particular** example of seeing whether or not a **particular** infinite set can be put into bijection with its power set. Afterwards, when we explore the case of a **general** infinite set, we will go back to writing A and $P(A)$.

subset $\{2, 4, 6, 8, 10, 12, ...\}$ in $P(\mathbb{N})$, the number 3 in $\mathbb{N}$ to the subset $\{1000001\}$ in $P(\mathbb{N})$, the number 4 in $\mathbb{N}$ to the subset $\{6, 7, 9327\}$ in $P(\mathbb{N})$, the number 5 in $\mathbb{N}$ to the subset $\{5, 10, 15, 20, 25, 30, 35, ...\}$ in $P(\mathbb{N})$, and so on, with infinitely many more assignments. We would then have an infinitely long list of assignments that F gives us, and it would then look like

$$1 \longrightarrow \{4, 12, 80\}$$
$$2 \longrightarrow \{2, 4, 6, 8, 10, 12, ...\}$$
$$3 \longrightarrow \{1000001\}$$
$$4 \longrightarrow \{6, 7, 9327\}$$
$$5 \longrightarrow \{5, 10, 15, 20, 25, 30, 35, ...\}$$
$$\vdots \quad \vdots$$

and the set S_1 would simply be $\{4, 12, 80\}$, S_2 would simply be $\{2, 4, 6, 8, 10, 12, ...\}$, and so on. However, in general we do not know the details of F — we are just **supposing** that such a bijective function exists — and therefore we must keep our notation completely general. This is why we are simply denoting the subset of $\mathbb{N}$ that F sends the element i to by S_i. And even though we do not know the details of each individual S_i, we will see shortly that we do not need to in order to arrive at a contradiction.

Using our infinitely long list of expressions of the form $F(i) = S_i$, we can construct a subset of $\mathbb{N}$ that is not in the list. This is in the exact same spirit of how we supposed that there was a bijective function from $\mathbb{N}$ to $\mathbb{R}$ and then used it to construct a real number that was not hit by this function. Namely, by constructing a subset which is not hit by this function F, we will have found the contradiction that we are after, since F is meant to hit **every** subset of $\mathbb{N}$ if it is truly bijective. We do this as follows.

Let us call this new subset T, and define it by the following. We take the element 1 in $\mathbb{N}$ and see if it is in S_1 (i.e, if 1 is in the subset which it maps to via the function F). Then, if 1 **is** in S_1, we **do not** put 1 in the set T, and if 1 **is not** in S_1, then we **do** put 1 in T. Thus, in our example above where $S_1 = \{4, 12, 80\}$, we **would** be placing 1 in T since 1 does **not** make an appearance in S_1.

We then move on to 2 in the set $\mathbb{N}$. If 2 is in S_2, then we do not put it in T, and if 2 is not in S_2, we do put it in T. In our example above where $S_2 = \{2, 4, 6, 8, ...\}$, we **would not** put 2 in the set T that we are constructing, since 2 **does** make an appearance in S_2. We then do this for all of the sets S_i. Namely, we define T to be the set which has the element i in it if and only if S_i does **not** have the element i in it. Therefore if S_i has the element i in it, then T does not, and if S_i does not have the element i in it, then T does.

This definition of T gives us a perfectly good subset of $\mathbb{N}$, and therefore T is an element of $P(\mathbb{N})$ since $P(\mathbb{N})$ is the set that has **all** of the subsets of $\mathbb{N}$ as its elements. Therefore T **should** be hit by our function F, since F is meant to be bijective and therefore surjective. But is it? Indeed, it is **not** hit by our function F because it differs from S_1 by either having or not having the element 1 (according to whether or not S_1 does). It also differs from S_2 in the element 2, and from S_3 in the element 3, and so on, all the way down the list. Indeed, if F **were** to hit our subset T, then there would be **some** element i in $\mathbb{N}$ such that $S_i = T$. However we know this is impossible, because only (and exactly) **one** of the two sets S_i and T contain the element i. Therefore T is **not** hit by our function F, and since this function was supposed to hit every element in $P(\mathbb{N})$, we clearly have a contradiction!

We have therefore found that no bijective function can be defined from $\mathbb{N}$ to $P(\mathbb{N})$, and thus $P(\mathbb{N})$ must have a fundamentally different cardinality. Namely, the cardinality of $P(\mathbb{N})$ must be a different infinity from that of $\mathbb{N}$, and indeed we know that the infinity of $P(\mathbb{N})$ must be **larger** than that of $\mathbb{N}$, since we can certainly define a **surjective** function from $P(\mathbb{N})$ to $\mathbb{N}$. Thus, by taking the power set of $\mathbb{N}$, we have jumped up to a new, larger infinity! Let us now see how we can use these ideas to generate infinitely many larger and larger infinities.

Infinity Infinities

Let us now suppose that A is any infinite set — any infinite set at all — and let us ask whether or not we can put A and $P(A)$ (the power set of whatever A is) into a bijective correspondence. The answer is no, and to show this we will use logic that is very similar to that which we used in the previous section to show that there is no bijection from $\mathbb{N}$ to $P(\mathbb{N})$.

As before, we begin by supposing that there **is** a bijective function F from A to $P(A)$. This means that for **any** element a in the set A, $F(a)$ is an element of $P(A)$ and therefore is a subset of A. We note that in the previous section when we took A to be $\mathbb{N}$, we knew that the elements of A (i.e., the elements of $\mathbb{N}$) were numbers. We then decided to denote $F(i)$ — the subset of $\mathbb{N}$ that the number i in $\mathbb{N}$ mapped to — by the symbol S_i. However, since A is now an **arbitrary** infinite set, it might not even be the case that its elements are numbers. For example, if $A = P(\mathbb{N})$, then A would indeed be an infinite set whose elements are **not** numbers, but rather **sets** of numbers. Thus, for any element a in the set A, we will simply denote by $F(a)$ the element of $P(A)$ that a maps to via the function F. In particular, we will not introduce the auxiliary notation "S_a," simply because we do not have to.[11]

We now define a new set T — a subset of A — as follows. For each element a in A, we put a into T if and only if a is not in $F(a)$. Namely, for every element a in A, we ask whether or not it is in the subset of A that it maps to via the function F, and we place a in T accordingly — if a is in $F(a)$, then we do **not** put a in T, and if a is **not** in $F(a)$, then we **do** put a in T. This defines a perfectly good subset of A and therefore a perfectly good element of $P(A)$.

However, by the same logic as above, this new subset T cannot possibly be mapped to by any element in A. This is because we have defined T to differ from $F(a)$, for any a in A. Namely, suppose there **is** an element b in A such that $F(b) = T$. We immediately know that this is impossible because only one of these two subsets of A have the element b in them, and thus they cannot be equal sets! Therefore T is a subset of $P(A)$ which is **not** mapped to by the function F, and this contradicts our supposition that F was bijective. Thus no bijective function from A to $P(A)$ can be defined, and so the cardinality of $P(A)$ must be a different type of infinity than that of A.

---

[11] Indeed, we did not have to introduce the extra notation "S_i" in the previous section either, but we did so anyway only because it is somehow more aesthetically pleasing to have numbers as subscripts.

Now, we again note that every element a in A makes up the subset $\{a\}$, which is an element of $P(A)$. We can therefore define a function from $P(A)$ to A which is **surjective**, by simply sending the 1-element subsets of $P(A)$ to their corresponding elements in A. We can then send all the other elements in $P(A)$ wherever we would like, since this function would already be surjective. Thus, since there is no bijective function from A to $P(A)$, and since there **is** a surjective from $P(A)$ to A, we know that the infinite cardinality of $P(A)$ is not only different from that of A, but also **larger** than it. Thus, for **any** infinite set A, the set $P(A)$ will have a truly **larger** infinite cardinality.

Now here is the catch — the punchline, so to speak. We have just shown that if A is any infinite set at all, then $P(A)$ is an infinite set with a truly larger cardinality. However, since $P(A)$ is itself an infinite set, the same logic that was used above can be used on $P(A)$ in place of A. Namely, there is nothing stopping us from taking the power set of the power set of some given set.[12] We would then know that the power set of $P(A)$, which we would denote[13] as $P(P(A))$, must have a larger (infinite) cardinality than $P(A)$. This is simply because $P(A)$ is some infinite set and we showed above that the power set of **any** infinite set has a larger cardinality than the set itself.

In the exact same way, $P(P(P(A)))$ (the power set of the power set of the power set of A) must have a larger infinite cardinality than that of $P(P(A))$, and $P(P(P(P(A))))$ must have an even larger infinite cardinality than $P(P(P(A)))$! We see that we can hit each successive power set with another "$P(\cdot)$" operation, thus uncovering another new, larger level of infinity. Moreover, we can do this forever, because each application of "take the power set of a given set" gives us back yet another infinite set with a larger infinite cardinality, and we can then take the power set of this new infinite set to get yet another infinite set with an even larger infinite cardinality, and so

---

[12] As an example, if $A = \mathbb{N}$, then the power set of $\mathbb{N}$ is $P(\mathbb{N})$ and its elements are subsets of $\mathbb{N}$, so that $\{1, 3, 12, 2056\}$ is one element in $P(\mathbb{N})$. Then, the power set of $P(\mathbb{N})$ is $P(P(\mathbb{N}))$, and its elements are subsets of $P(\mathbb{N})$. Thus, $\{\{1, 2, 5\}, \{7\}, \{2, 4, 6, 8, 10, ...\}, \{1017, 4056\}\}$ is **one element** of $P(P(\mathbb{N}))$.

[13] Recall that notation is nothing but some abstract instructions, and the notation "$P(\cdot)$" is simply instructions for "take the power set of whatever replaces the "$\cdot$" on the inside of the parentheses." Thus, $P(P(A))$ means to take the power set of $P(A)$, which itself is the power set of A.

on. Namely, if the enemy hands us some set, let us call it B, and says that it is the "most infinite" set out there, then we can immediately tell him he is wrong because we know that $P(B)$ is fundamentally more infinite[14] than B, and that $P(P(B))$ is even more infinite than that, and so on. Thus, not only is the set that the enemy handed us **not** the most infinite, there are **infinitely many** larger infinities!

This is identical to why the enemy can never hand us the "largest" finite number. If the enemy handed us a finite number and said that it was the largest finite number, we would immediately know that he was wrong because we can add 1 to it and get a larger number. Indeed, we could even add 1 to the number we just got to get an **even larger** number, and so on. In this way, we see that the number the enemy handed us is not only not the largest number, but indeed has **infinitely many** larger numbers sitting above it. Thus, by showing that the power set of **any** infinite set has a larger cardinality than the set itself, we have found a way to literally "add 1" to any infinite number[15] in order to get a different, truly larger infinite number. And indeed, since we can "add 1" to infinity as many times as we would like, this tower of infinities is infinitely high!

Parting Thoughts On Infinity

Before ending this chapter and moving on to wildly different concepts for the rest of this volume, let us introduce some of the ideas and unsolved problems that exist out there in the world of math. We will discuss these more in future volumes but it does not hurt to introduce them now. And since this is just an introduction, we will not discuss these ideas in depth and therefore some confusion may arise. This is okay, though, as it will encourage us to keep moving forward into this amazing world of mathematics.

In the previous section of this chapter we saw a way of "adding 1" to infinity, in order to get another, larger infinity. We compared the procedure of taking the power set of an infinite set to the procedure

---

[14]In the very well-defined sense of having a truly larger cardinality.

[15]We note that this "adding 1" operation is most certainly **not** simply adding a **single** element to our infinite set, but rather "adding 1" to the **cardinality itself**, which is achieved by taking the power set of our infinite set.

of adding 1 to a normal, finite number. The natural next question to ask, then, is whether or not there is an analog to adding one half to a normal, finite number. Namely, we know that we can add 1 to 1 to get 2, but we also know that if we allow for fractions then we can add $\frac{1}{2}$ to 1 to get $\frac{3}{2}$. The result of adding $\frac{1}{2}$ to a number is to get a number that is truly **in between** the number itself, and the number that is obtained by adding 1. Namely, $\frac{3}{2}$ is truly in between 1 and 2, because it is greater than 1 and less than 2. Thus, we might ask whether or not there are any infinite sets that have a truly larger cardinality than $\mathbb{N}$ and a truly smaller cardinality than $P(\mathbb{N})$.

For this to be the case, we would need a set A such that there is no bijective function from A to **either** $\mathbb{N}$ or $P(\mathbb{N})$, so that the cardinality of A does not equal the cardinality of either $\mathbb{N}$ or $P(\mathbb{N})$. Additionally, this set A would need to be such that there was a **surjective** function from A to $\mathbb{N}$ (so that the cardinality of A is truly larger than that of $\mathbb{N}$), **as well as** a surjective from $P(\mathbb{N})$ to A (so that the cardinality of A is truly smaller than that of $P(\mathbb{N})$). If we could find such a set, then we will have found an infinity that is "between" the infinity of $\mathbb{N}$ and that of $P(\mathbb{N})$. One natural guess for such a set would be $\mathbb{R}$, since we already know that it has a larger cardinality than $\mathbb{N}$. Therefore, if we could show that $P(\mathbb{N})$ has a larger cardinality than $\mathbb{R}$ we will have solved the problem, and the answer would be that there does exist a set whose cardinality is between $\mathbb{N}$ and $P(\mathbb{N})$. It turns out, however, that the cardinality of $\mathbb{R}$ is **the same** as that of $P(\mathbb{N})$, though we will not provide the proof of this statement here.[16] This means, however, that we have not yet found a set whose cardinality is between that of $\mathbb{N}$ and $P(\mathbb{N})$. Let us therefore ask: Is there such a set at all?

This is a perfectly reasonable question to ask, but it turns out that no one knows the answer to it. Georg Cantor — the brilliant mathematician who developed this entire formalism regarding infinity — died before he could figure it out, and every other mathematician that has tried to figure it out is either still struggling over it or has also died before succeeding. Namely, there is a statement called the Continuum Hypothesis which states that there is **not** a set whose car-

---

[16]The proof relies on some pretty heavy machinery that we have not yet developed here, but the reader is certainly encouraged to try to prove this herself, or at least convince herself of the truth of this statement.

dinality is between that of ℕ and that of $P(\mathbb{N})$, but no mathematician has been able to **prove** that this is a true statement. Indeed, the Continuum Hypothesis is unlike many other unsolved problems in math — the problem **itself** has a problem. It has in fact been shown that, using our standard system of mathematical axioms, the Continuum Hypothesis is actually **unsolvable**. What we mean by "unsolvable" is that even if it is true, we cannot **prove** that it is so. The possibility of fundamentally not being able to prove something within a given set of axioms comes from the remarkable result uncovered by the mathematician Kurt Gödel, who proved what is known as the Incompleteness Theorem.[17] The Incompleteness Theorem effectively tells us that in order to **prove** that the statement "there is no set whose cardinality is strictly larger than that of ℕ and strictly smaller than that of $P(\mathbb{N})$" is true, we will have to change the very foundations of mathematics! This is most certainly not a small task to ask for, but the difficulty of a problem has never stopped mathematicians in the past!

Before ending this chapter we should note how all of this discussion about infinity is only the tip of a very large iceberg. To see this, let us recall yet again how we have constantly used various properties of the finite numbers to motivate various definitions in the infinite case. We saw that this analogy allowed us to define some very abstract notions of counting, and these definitions allowed us to uncover a remarkable organization and order within the realm of the infinite. It turns out that we can still do a lot more. Namely, there is a beautiful and abstract way to define addition, subtraction, multiplication, division, and exponentiation in such a way as to simultaneously incorporate the finite, as well as the infinite numbers. In doing so, we will have put all of the various infinities on essentially equal footing with all of the various finite numbers, thus extending our very notion of what a number is to a much larger class of numbers — a class of numbers which includes the previously mysterious notion of infinity.

In order to see all of this — i.e., the details of the Continuum Hypothesis, the implications of the Incompleteness Theorem, and the ways in which we can add, subtract, multiply, and divide infinite

---

[17] We will have more to say about the Incompleteness Theorem and its profound impact on math in a future volume.

numbers just like finite ones — we must develop some more mathematical machinery. Along the way, we will see lots of other forms of mathematical beauty over a wide range of topics, so let us temporarily leave the world of the infinite behind us and move forward to develop more mathematical weaponry.

Chapter 16

Unions And Intersections

How We Approach Mathematical Structures

We are now going to shift gears entirely from our discussion of the various kinds of infinities that exist and return to our more general discussion of sets. Let us therefore recall what we had done before we went on our detour in Chapter 7 into the wonderful world of the infinite. We had defined sets in Chapter 1 and then we realized that a natural idea to want to develop is a way to "relate" one set to another. In other words, once we defined what a set is, we wanted to know about the kinds of things that happened when we have two sets lying around. This motivated our definition of a function in Chapter 6, which gives us a precise way of speaking about "associations" between the elements of two sets. Another natural question to ask about sets is what sort of "sub-structure" they have. In other words, if the enemy hands us a set, we might want to know if there are meaningful ways for us to "deconstruct" it into "smaller" things? This motivated, in a natural way, our definition of a subset in Chapter 2.

These two natural questions about sets — namely, how to define associations **between** sets and how to define sub-structure within a **single** set — carry over to almost all mathematical structures that one can define. As we will continue to see throughout these volumes, this gives us a very direct method of moving forward in mathematics.

HOW WE APPROACH MATHEMATICAL STRUCTURES 161

In particular, once we define some mathematical structure[1] we can ask about what happens when two (or more) of such structures are around. This will give rise to whatever the analog of a function is with regard to this new mathematical structure. Moreover, we can also ask what kind of "sub-structures" exist within any such structure. This would be the analog of a subset. We recall that an important quality of a subset is that it is **itself** a set in its own right. In other words, a subset of a set has the same structure (namely, the structure of being a set) that its "parent" set[2] has. Thus, for any mathematical structure that we define, it is important to know if it allows some kind of sub-structure that mirrors the larger, "parent" structure.

We will see more examples of this when we introduce mathematical structures such as groups, topological spaces, and categories, which are all **much** more complex structures than sets. Accordingly, the analogs to the "functions" between such structures and the "subsets" of these structures will be more fascinating as well. We will not be exploring these other structures for quite a while, but it is important to keep in mind the generality of our line of inquiry — namely, define a structure, ask about relationships **between** such structures, and ask about the sub-structure of these structures — as this type of inquiry is incredibly useful regardless of which structure we are exploring at any given time.

Another very natural — and very general — question to ask about a given mathematical structure is whether or not we can "build" new structures out of old ones. In other words, if we are handed two of these structures, can we define another of these structures **using** those that we were given. The answer is almost always yes, and in this chapter we will study this question as it pertains to sets.

Before doing so, however, let us note that this will in some sense complete the picture of "natural" questions[3] to ask about a given structure. In particular, once we see how to build a new structure

---

[1] In our case so far, this mathematical structure has simply been a set.

[2] By "parent" set we simply mean the set that the subset is a subset of. Thus if B is a subset of A, then we can refer to A as the parent set of B. This is not the technical terminology, it is just useful and intuitive.

[3] We keep "natural" in scare quotes because by "natural" we really mean "obvious" and "necessary," in the sense that we **need** to ask these questions if we are to ever make progress in studying these structures.

from two old ones, we will then know how to relate two structures to each other (the analog of the function), find smaller structures within a structure (the analog of a subset), and build up new structures from other structures (the analog of what we are going to do in this and the next couple of chapters). In other words, we will know how to relate structures, decompose structures, and build up structures. Once we have all three of these tools at our disposal, we can begin to make progress studying these structures.

Two Important Constructions

Now that we have all of this philosophizing behind us, we can get on with mathematics and see how this general approach pertains to sets. We therefore ask, given two sets A and B, what kind of sets can we create **using** A and B? I.e., how can we build a new set from the two old sets A and B?

There are several definitions that we can make — namely, there are several different ways that we can take two sets A and B and form a third set — but the two that we will study here are **unions** and **intersections** of sets. In short, the **union** of two sets A and B is the set that is formed by "bringing these sets together," and the intersection of two sets is formed by considering the parts of the sets that are "the same." Let us make this a bit more rigorous.

Definition 16.1. Given two sets A and B, the **union** of A and B is the set that consists of the elements which are **either** in A **or** in B. The union of A and B is denoted by $A \cup B$.

Note that we should really view $A \cup B$ as a **single** symbol representing the **single** set called "the union of A and B," in the exact same way that, for example, "A" is a single symbol representing whatever the set A is. Thus, any element in $A \cup B$ is either an element of A or an element of B (or possibly both), and conversely any element in A or B is also an element of $A \cup B$. For example, if we are handed the set $A = \{1, 2, 3\}$ and the set $B = \{3, 4, 5\}$, then the **union** of A and B, which is denoted by $A \cup B$, is the single set given by $A \cup B = \{1, 2, 3, 4, 5\}$. Note that we do not need to count the element 3 twice simply because, by definition, sets only consist of

distinguishable elements, and the 3 that appears in A is identical to that which appears in B. We note that indeed, as the definition requires, every element in $\{1, 2, 3, 4, 5\}$ is **either** in A **or** in B (or in both). As a second example, let us suppose now that $A = \{1, 2, 3\}$ still, but that $B = \{a, b\}$. We then have that $A \cup B = \{1, 2, 3, a, b\}$. Thus, constructing the union of two sets is indeed one way of taking two "old sets" (in this case A and B) and forming a "new" set (in this case, $A \cup B$).

Let us now take this opportunity to introduce some more notation that we will end up using quite a bit. For most of this notation, all we are doing is introducing symbols to abbreviate terminology that we have already been using for quite a while, so there really is not anything new happening here. To begin, let us agree to denote the statement that some element, call it a, is in a set A, by the symbol "$\in$." In particular, let us agree that when we write $a \in A$, we should read it as "a is an element of A." Thus, whatever is on the "back side" of $\in$ is an element, and whatever goes on the side that the three prongs of $\in$ point to is the set that the element is in. For example, if $A = \{1, 2, 3\}$, then the statement "$1 \in A$" would be true. As another example, suppose $B = \{2, 4, 6, 8\}$. We can then truthfully say "for any $b \in B$, B is even," simply because this sentence reads "for any element b in the set B, b is even," and this is indeed a true statement.

Let us now introduce another bit of notation which is extremely useful for defining sets. Namely, let us suppose we wanted to define a set A by saying that all of the elements in A satisfy some condition. For example, we could define the set that consists of all the prime numbers that are greater than 50 and less than 50,000. Constantly referring to "the set of prime numbers greater than 50 and less than 50,000" is tiring, and extremely annoying to write down on paper. Therefore, let us agree to streamline the statement that "A is the set of prime numbers greater than 50 and less than 50,000," by writing

$$A = \{p \in \mathbb{N} \mid p \text{ is prime}, 50 < p < 50,000\}.$$

As usual, we are using the bracket notation $\{\}$ to surround the elements of the set, but now we are actually **describing** the elements in the set within these brackets. On the left of the vertical line "|" we write which type of objects we are dealing with — in this case these are elements in ℕ. Then on the right side of the vertical line we place the desired conditions on the elements in ℕ. The vertical line should be read as "such that," so that the whole equation reads, from left to right, "A is the set of elements p in the set ℕ such that p is prime, and is greater than 50 and less than 50,000." The only way to get used to this notation is to see some more examples, so let us therefore now take an example that is appropriate for this chapter.

Namely, using this notation we can write the union $A \cup B$ of two sets A and B as follows:

$$A \cup B = \{a \mid a \in A \text{ or } a \in B\}.$$

This reads as, from left to right, "the union of A and B is the set of elements a such that a is in A, or a is in B." It is important to note that this statement, when written out in symbols as it is above, is genuinely grammatically correct. It is always in good taste to use our mathematical symbols in as grammatically correct of a way as possible.

With this notation in hand, we can now define the second construction of sets that we will see in this chapter — namely, the intersection of two sets — very easily.

Definition 16.2. Given two sets A and B, the **intersection** of A and B is the set of elements that are in **both** A **and** B. We denote the intersection of A and B by $A \cap B$, and we therefore see that $A \cap B = \{a \mid a \in A \text{ and } a \in B\}.$

Let us note that, in words, the mathematical expression

$$A \cap B = \{a \mid a \in A \text{ and } a \in B\}$$

reads as "the intersection of A and B is the set of all elements a such that a is in a **and** a is in B." As usual, we should look at a couple easy examples in order to get used to this definition.[4] Suppose $A = \{1, 2, 3\}$ and $B = \{3, 4, 5\}$, so that this is the same as the first

---

[4]As a side note, one easy way to remember that "∪" is the symbol corresponding to the union of two sets while "∩" is the symbol corresponding to the intersection of two sets is to note that "∪" looks like a "U," and "union" starts with a "U." Then, the intersection symbol is simply the one that is not "∪."

example that we looked at after defining unions above. We then have that $A \cap B$ is the set of elements that are **both** in A and B, so that we have $A \cap B = \{3\}$, a 1-element set. As a second example, let us now suppose that $A = \{1, 2, 3\}$ and $B = \{a, b\}$. We saw above that the **union** of these two sets is $A \cup B = \{1, 2, 3, a, b\}$, however now we see that the **intersection** of these two sets is **empty**. We can write this as $A \cap B = \emptyset$, and the reason why $A \cap B$ is empty is simply that there is no element that is in both A and B.

There is a quick and easy way to visualize unions and intersections of sets using the Venn diagrams in Figures 16.1 and 16.2.

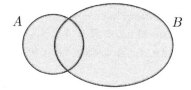

Figure 16.1: A visualization of the union of A and B in terms of a Venn diagram.

Namely, if we represent the set A by the circle on the left of these figures (where the elements in A are meant to, in some sense, lie within its corresponding circle) and if we represent the set B by the ellipse on the right of these figures, then the union of these two sets is the set obtained by "filling in" both of the sets. Similarly, the intersection of these two sets is the set obtained by "filling in" only the parts of the two sets that overlap, because this is the region that represents those elements that are in **both** A and B.

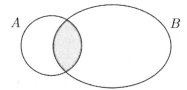

Figure 16.2: A visualization of the intersection of A and B in terms of a Venn diagram.

We will see more examples of unions and intersections of sets in the exercises at the end of this chapter, so for now let us make a couple of quick, general remarks. The first remark we should make may seem somewhat pedantic or unimportant, but it is indeed extremely important for the overall logical structure of mathematical arguments. Namely, we need to be more careful with how we use our "=" symbols. The reason for this is that — in a very significant manner — we have been using our "=" symbol in two distinct ways so far, and these two different uses need to be distinguished.

In particular, it is useful to distinguish when two things are equal because they are **defined** to be equal, and when two things are equal because they **turn out** to be equal. For example, we can **define** A to be the set of all whole numbers between the number 5 and the number 7, and we can also **define** the set B to be the set of all numbers that are equal to 3 times 2. In the notation that we introduced above, we have

$$A \text{ "is defined to be" } \{p \in \mathbb{N} \mid 5 < p < 7\}$$
$$B \text{ "is defined to be" } \{p \in \mathbb{N} \mid p = 3 \times 2\}.$$

Now, it **turns out** that the only whole number between 5 and 7 is the number 6, so that $A = \{6\}$. It also **turns out** that the only number that equals 3 times 2 is 6, so that $B = \{6\}$. Thus, it **turns out** that $A = B$, but it is **not** the case that A **is defined to be** B. We therefore have two distinct types of equalities. One is when two things are equal because they are simply defined to be equal, and the other is when two things just happen to be equal due to some logical manipulations of our definitions. It is important for us to distinguish these two logically distinct phenomena, so let us agree to use the usual "=" sign when things **turn out** to be equal, and let us agree[5] to use "≡" when one thing is defined to be another. Thus, in the above example, we have that $A \equiv \{p \in \mathbb{N} \mid 5 < p < 7\}$ and $B \equiv \{p \in \mathbb{N} \mid p = 3 \times 2\}$, and that after a little thought we see that $A = B$. To be totally precise, we should have been using this

---

[5] We should note that many mathematicians and textbooks also use the ":=" sign in place of the "≡" sign. This is nothing but a notational convention that mathematicians have yet to fully agree on, so that ":=" and "≡" represent **precisely** the same ideas. Accordingly, we may bounce back and forth as well just to get used to both sets of notation.

TWO IMPORTANT CONSTRUCTIONS 167

distinction in all of the above definitions that we have made, so that for example $A \cap B \equiv \{a \mid a \in A \text{ and } a \in B\}$. However, we can allow some imprecision with this notation **only if** it is clear what a given equality sign means (i.e., whether it represents a definition or a result of some logical argument). This might seem a bit pedantic, but it is actually a very important tool for keeping the logic within any given argument organized, and we will see this distinction pop up more as we go.

The final remark that we will make in this chapter is on the importance of the empty set. It is clear that if **neither** A nor B are the empty set, then the union $A \cup B$ will also not be the empty set. However, even if A is the empty set, we have that $A \cup B$ will not be empty so long as B is not empty. This is simply because of the fact that the union of two sets is the set of elements that are in **either** of the two sets that we are taking the union of. Thus, if B is not empty, then all of its elements will be in $A \cup B$ and therefore $A \cup B$ will not be empty either.[6] However, it might be the case that the intersection of two sets is empty even if both of the sets that we are taking the intersection of are **not** empty. We saw an example of this above, where we took $A = \{1, 2, 3\}$ and $B = \{a, b\}$. In this case, neither A nor B are empty, yet their intersection is indeed empty. As another example, the intersection of the sets

$$\{\text{people who say Lebron is better than Kobe}\}$$

and

$$\{\text{people who have never been wrong}\}$$

is empty, because there is no element that is in both of these sets. The fact that two non-empty sets can have an empty intersection is not only an example of how different unions and intersections are, but also a reflection of the fact that the empty set truly exists. Namely, we really need to incorporate the set with no elements into our mathematical formalism, otherwise many of our definitions will not make

---

[6]It is important to note, though, that if A and B are **both** empty, then $A \cup B$ is empty as well. This can be seen by noting that if there are no elements in A and no elements in B, then there cannot be any elements in the set of elements that are either in A or B.

sense. If we were not able to use the empty set as a genuine set, then we would not be able to define intersections at all, since most pairs of sets have empty intersections. Indeed, we will see **many** instances throughout these volumes in which the existence of the empty set is truly vital.

The following exercises are extremely important for solidifying our knowledge of unions and intersections, which are two of the most fundamental and foundational constructions that we can make with sets. These constructions will appear over and over again in all of mathematics, so it is important that we understand them fully.

Exercises

1) How many elements are in the union of the sets $\{1, 2, 3, 4, 7\}$ and $\{3, 4, 5\}$? How many are in their intersection?

2) What is the union of the sets

$$\{p \in \mathbb{N} \mid p \text{ is even}\}$$

and
$$\{p \in \mathbb{N} \mid p \text{ is a multiple of } 4\}?$$

What is their intersection?

3) What is the intersection of the sets $\{a \in \mathbb{N} \mid a \text{ is odd}\}$ and $\{a \in \mathbb{N} \mid 4 < a < 12\}$?

4) Let A be any set at all, and let O be a subset of A. What is $A \cup O$? What is $A \cap O$?

5) Let A and B be any sets at all. Prove that A and B are both subsets of $A \cup B$. Also prove that $A \cap B$ is a subset of A as well as a subset of B.

6) Let A be any set at all. What is the union of A with the empty set? What is their intersection?

Chapter 17

Playing Well With Others

A Whole New Class Of Questions

In this chapter we are going to ask and answer a class of questions that are absolutely pervasive throughout all of mathematics, in the sense that we will use the results derived here over and over again throughout our **entire** relationship with math. Accordingly, the importance of the results that we will establish in this chapter simply cannot be overstated. Additionally, the significance of the **nature** of these questions cannot be overstated either, for we will end up asking analogous questions in every single area of math that we end up devoting our attention to. This is an extremely important aspect of mathematics, and one that we have tried and will continue to try to keep aware of — namely, we should not only focus on the questions and their answers in an isolated manner, but rather we should always be aware of the significance of the questions in a larger and more general setting. By doing so, we will be much more in tune with the beautiful unity that seemingly disparate fields of math have to offer.

Accordingly, let us set the stage for how all of this philosophizing applies to our current situation. We must first recall that our definition of elements and sets back in Chapter 1 is what got the ball rolling. Once we were able to define elements and sets, we were then able to start defining lots of other structures that manipulated them. Namely, we defined **subsets** of sets, functions **between** sets, and ways to build **new** sets from old ones by taking the union or intersection of two sets to form another set. We have already remarked on the generality of this procedure — i.e., any mathematical structure that we will define will end up having a natural notion of "sub" structure, a natural analog of functions between structures, and various natural ways to build up new structures from old ones. What we are now tasked with is answering the brand new class of questions pertaining to how these various ways of manipulating mathematical structures **play with each other**.

There is a somewhat precise definition of what we mean by having two types of manipulations on some mathematical structure "play well with each other," but we will not be able to make this definition until a later volume. We can, however, get a sense of what this means by considering explicit instances in which different manipulations play with each other. It will then begin to become clear what is meant by having manipulations play **well** each other, as opposed to the alternative (i.e., not well).

For example, we have seen that if we have two sets A and B, then we can take the union of them to form the single set $A \cup B$. Now let us suppose that we have a third set, C. There are then several questions that we can ask. In particular, if we take the union of A and B to form the set $A \cup B$, then there is nothing stopping us from taking the two sets $A \cup B$ and C and forming **their** union, thus creating the set[1] $(A \cup B) \cup C$. However, we very well could have **first** taken the union of, say, B and C to form the set $B \cup C$ and then taken the union of A and $B \cup C$ to form the set $A \cup (B \cup C)$. We can then ask the question of how $(A \cup B) \cup C$ and $A \cup (B \cup C)$ are related to each other — if they are nicely related to each other then we can think of the procedure of taking a union as "playing well with itself," otherwise we can think of this procedure as not playing well with itself. There are tons of more complicated questions that we can

---

[1] We note that the parentheses in the following expression are extremely important for the logical structure of our argument. These parentheses denote the fact that we are **first** taking the union of A and B, and **then** taking the union of the single set $A \cup B$ with the set C. This is one place where it is extremely important to recall that "$A \cup B$" is truly a symbol representing **one** set — the union of A and B.

ask about how unions and intersections interact, or play with each other, and establishing these results will be the subject of the next sections of this chapter.

Unions Of Unions

We begin by answering the question that we posed in the previous section. Namely, suppose we now have three sets A, B, and C. We can then take the union of A and B to form the set $A \cup B$, and then we can form the union of $A \cup B$ with C to create the set $(A \cup B) \cup C$. However, we can also first take the union of B and C to form $B \cup C$ and then take the union of this set with A to form $A \cup (B \cup C)$. We now want to ask whether or not these two sets are equal. Namely, we want to know if

$$(A \cup B) \cup C \stackrel{?}{=} A \cup (B \cup C).$$

Before answering this let us also revisit the question of when two sets are equal in general, and in doing so we will introduce a powerful method for proving that two sets are equal.

Let us recall that two sets A and B are equal (by definition) when their elements are identical. One way to phrase this is that two sets are equal if there is no element that is in one set but not in the other set. Yet another way to phrase this is that two sets are equal only if every element that is in one set is in the other set, **and conversely**. Namely, two sets A and B are equal if every element that is in A is also in B, **and** every element that is in B is also in A. This is nothing but a restatement of the fact that the elements in A are exactly the same as the elements in B.

If we reflect on this way of phrasing things — namely that two sets A and B are equal if every element in A is in B and if every element in B is in A — then we see that this is nothing but the statement that A and B are equal **if they are both subsets of each other**. This is because the fact that "every element in A is in B" is nothing but the statement that A is a subset of B. Additionally, the statement that "every element in B is in A" is nothing but the statement that B is a subset of A. We therefore see that two sets are equal if and only if they are both subsets of each other. Let us introduce some notation that will help us abbreviate all of this.

Let A and B be two sets and suppose that B is a subset of A. Let us agree to denote this fact by $B \subseteq A$. The symbol "$\subseteq$" is analogous to the symbol "$\leq$" used for numbers, where we denote the fact that a number b is less than or equal to a number a by writing $b \leq a$. Thus, in a very vague sense, we can view the notation "$B \subseteq A$" as denoting the fact that B is "less than or equal to" A. This is only an analogy, however, and the **precise** meaning of the notation "$B \subseteq A$" is that every element that is in B is also in A.

We now recall that a number b is less than or equal to a number a, if and only if it is also the case that a is greater than or equal to b. Therefore, writing $b \leq a$ has exactly the same logical content as writing $a \geq b$. Using this as motivation, let us also introduce the symbol "$\supseteq$," so that we can equally well denote the fact that B is a subset of A by the notation $A \supseteq B$. This is nothing but different notation for the exact same logical statement, so that $B \subseteq A$ and $A \supseteq B$ mean the exact same thing.[2]

Our discussion above then tells us that two sets A and B are equal if and only if $B \subseteq A$ **and** $A \subseteq B$. We note that this is also analogous to the case with numbers. Namely, the only way for two numbers a and b to be such that $b \leq a$ **and** $a \leq b$ is if $a = b$. In other words, two numbers are **both** less than or equal to each other if and only if they are exactly equal to each other.[3] Similarly, two **sets** are both subsets of each other if and only if they are equal to each other.

The above might all seem a little unnecessary, but it actually shows us a powerful method for determining if two sets are equal to each other. Namely, we can show that $A = B$ by simply showing two easier statements — that $A \subseteq B$ and that $B \subseteq A$. And in order to show that one set is a subset of another, we need to show that every element in the former set is also in the latter set. From the notation that we introduced in the last chapter, this means that if we want to show that B is a subset of A, then we need to show that if $b \in B$,

---

[2] Note that in both cases the "open" side of "$\subseteq$" is pointing to the "larger," or "parent" set.

[3] For example, 4 is less than or equal to 5, but 5 **is not** less than or equal to 4. This is what we expect, because 5 and 4 are not equal to each other.

UNIONS OF UNIONS

then it is also true that $b \in A$. In other words, any element b that is in B is also in A.

Before actually proving some results, let us recall some notation that will greatly increase our mathematical sophistication. We just discussed how in order to show that B is a subset of A, we need to show that any element $b \in B$ is also in A. Our future proofs will be streamlined by recalling[4] that the notation "$\Rightarrow$" simply means "implies." As an example of using this notation, we can write that if we want to show that $B \subseteq A$, then we need to show that

$$b \in B \Rightarrow b \in A.$$

This is nothing but fancy notation for the statement that "if b is an element in B then that implies that b is an element of A," and this is precisely what we want to show in order to establish that B is a subset of A. Namely, if it is true that b being in B implies that b is in A, then B is a subset of A. Therefore, in order to show that $A = B$, we need to show that A and B are both subsets of each other, and we have just seen that one way to do this is to show that

$$a \in B \Rightarrow a \in A$$

and that
$$a \in A \Rightarrow a \in B.$$

Let us now return to our initial question and use all of this fancy new notation.

It turns out that indeed, for **any** three sets A, B, and C, the following is true:

$$(A \cup B) \cup C = A \cup (B \cup C).$$

In order to prove this, we need to show that $(A \cup B) \cup C \subseteq A \cup (B \cup C)$ and that $A \cup (B \cup C) \subseteq (A \cup B) \cup C$. Let us not be intimidated by these two expressions, for the statement "$(A \cup B) \cup C \subseteq A \cup (B \cup C)$" simply says that "$(A \cup B) \cup C$ is a subset of $A \cup (B \cup C)$," and similarly for the expression "$A \cup (B \cup C) \subseteq (A \cup B) \cup C$." We will begin by

---
[4]See Chapter 15 for where this notation was initially introduced (and therefore more fully explained).

showing that $(A \cup B) \cup C \subseteq A \cup (B \cup C)$. We first need to take any arbitrary element of $(A \cup B) \cup C$ and see if it is also in $A \cup (B \cup C)$. Accordingly, let a be some arbitrary element in $(A \cup B) \cup C$. By definition of what the union of two sets means, we have that

$$a \in (A \cup B) \cup C \Rightarrow a \in A \cup B \text{ or } a \in C.$$

In words, this simply means that "if a is an element of $(A \cup B) \cup C$, then it is an element of $A \cup B$ or it is an element of C (or both)," and this is true simply because the union of $A \cup B$ and C is **by definition** the set of elements that are either in $A \cup B$ or in C. Now, if $a \in A \cup B$, then a is either in A or in B, by definition of the set $A \cup B$. We therefore have, adding another line to our chain of logic above, that

$$a \in (A \cup B) \cup C \Rightarrow a \in A \cup B \text{ or } a \in C$$
$$\Rightarrow a \in A \text{ or } a \in B \text{ or } a \in C.$$

Now, if $a \in B$ or $a \in C$, then we can write that $a \in B \cup C$ because this is again the definition of the set $B \cup C$. We can now add yet another line to our chain of logic and write

$$a \in (A \cup B) \cup C \Rightarrow a \in A \cup B \text{ or } a \in C$$
$$\Rightarrow a \in A \text{ or } a \in B \text{ or } a \in C$$
$$\Rightarrow a \in A \text{ or } a \in B \cup C.$$

Finally, we see that if $a \in A$ or $a \in B \cup C$, then this is nothing but the definition of a being in the set $A \cup (B \cup C)$. We therefore can add our last line in this chain and write

$$a \in (A \cup B) \cup C \Rightarrow a \in A \cup B \text{ or } a \in C$$
$$\Rightarrow a \in A \text{ or } a \in B \text{ or } a \in C$$
$$\Rightarrow a \in A \text{ or } a \in B \cup C$$
$$\Rightarrow a \in A \cup (B \cup C).$$

This is precisely what we wanted to show — namely, that every element in $(A \cup B) \cup C$ is also an element of $A \cup (B \cup C)$, and we have therefore shown that $(A \cup B) \cup C \subseteq A \cup (B \cup C)$. We note that

UNIONS OF UNIONS

we did this using nothing but the definition of the union of two sets at each stage of the proof, and therefore this is a completely general result for the union of **any** three sets.

We now just need to show that $A \cup (B \cup C)$ is also a subset of $(A \cup B) \cup C$ in order to finish our proof that $(A \cup B) \cup C = A \cup (B \cup C)$. The chain of logic that proves this is almost identical to that which we used above, so let us just state it using our fancy new notation. We have that

$$a \in A \cup (B \cup C) \Rightarrow a \in A \text{ or } a \in B \cup C$$
$$\Rightarrow a \in A \text{ or } a \in B \text{ or } a \in C$$
$$\Rightarrow a \in A \cup B \text{ or } a \in C$$
$$\Rightarrow a \in (A \cup B) \cup C.$$

It is enlightening to remember what all of this notation means in words, so let us write this whole chain out in all of its gory detail.[5] We have that "if a is an element of $A \cup (B \cup C)$ then this implies that a is an element of A or an element of $B \cup C$ and this implies that a is an element of A or of B or of C and this implies that a is an element of $A \cup B$ or of C and this implies that a is an element of $(A \cup B) \cup C$." This proves that $A \cup (B \cup C) \subseteq (A \cup B) \cup C$, and when we combine this with the previous result that $(A \cup B) \cup C \subseteq A \cup (B \cup C)$, we see that we have now **proved** that $(A \cup B) \cup C = A \cup (B \cup C)$.

In order to clarify this whole discussion, let us show an explicit example of the statement $(A \cup B) \cup C = A \cup (B \cup C)$. Namely, let us suppose $A = \{1, 2, 3, 4\}$, $B = \{2, 4, 6, 8\}$, and $C = \{2, 6, 7, 8\}$. In order to construct the set $(A \cup B) \cup C$, we need to first take the union of A and B. We quickly see that we have $A \cup B = \{1, 2, 3, 4, 6, 8\}$. We now need to take the union of $A \cup B$ with the set C, and since the only element that is in C that is not already in $A \cup B$ is 7, we see that we only need to add 7 to our set. Namely, we have that $(A \cup B) \cup C = \{1, 2, 3, 4, 6, 7, 8\}$. Now, in order to construct the set $A \cup (B \cup C)$, we need to first take the union of B and C. We quickly see that $B \cup C = \{2, 4, 6, 7, 8\}$, and once we take the union of this set with A we see that $A \cup (B \cup C) = \{1, 2, 3, 4, 6, 7, 8\}$. And indeed, we

---
[5] The length of the following run-on sentence will likely make our high school English teachers cry.

see that the two sets $(A \cup B) \cup C$ and $A \cup (B \cup C)$ are equal, as they should be.

The point of all of this was not only to start to get used to our new notation, but also to show that **the order in which we take the union of three sets does not matter** — the elements of the resulting set will be exactly the same regardless of which two sets we take the union of first. We can therefore be lazy with our parentheses and simply write $A \cup B \cup C$ (without any parentheses at all), since we now know that it does not matter which pairs of sets we choose to take the union of first. Indeed, we now know that we can just view the set $A \cup B \cup C$ as being the set of all elements from any of the three sets, which is what we want from our notion of "union-ing" sets. In regards to the title of this chapter, we have now **proven** that the process of taking the union of sets plays well with itself.[6] Let us now go on to see how intersections of multiple sets behave, as well as how unions of intersections (and intersections of unions) behave. Our new notation will streamline this discussion greatly.

Intersections Of Intersections

Let us dive right in to proving more results about what happens when we take unions and intersections of more than two sets. In the previous section we considered the sets $(A \cup B) \cup C$ and $A \cup (B \cup C)$ for any three sets A, B, and C, and we saw that these two "unions of unions" are equal. Let us therefore go on to consider $(A \cap B) \cap C$ and $A \cap (B \cap C)$ for any three sets A, B, and C. Namely, let us see if the order in which we intersect any three sets matters. Indeed, we will find that the order does not matter, and therefore that if we first intersect A and B and then intersect this with C then we get the same result as if we first intersect B and C and then intersect this with A.

To do this, we use the same method that we used in the previous section. Namely, we will show that $(A \cap B) \cap C$ is a subset of $A \cap (B \cap C)$, and vice versa. Thus, we need to show that if a is any element

---

[6]For, as we will continue to see, the definition of a union would **not** be very useful if it mattered which pair of sets we took the union of first whenever we wanted to take the union of multiple sets.

in $(A \cap B) \cap C$ then it must also be an element of $A \cap (B \cap C)$, and conversely that if a is any element in $A \cap (B \cap C)$ then it must also be an element of $(A \cap B) \cap C$.

The logic ends up being exactly the same as in the case of unions, only with the word "or" replaced with "and." Namely, if a is an element in $(A \cap B) \cap C$, then it must (by definition of the intersection of two sets) be in the set $A \cap B$ **and** in the set C. But if a is in the set $A \cap B$ then it must be in the set A **and** the set B. Thus a must be in A and B and C. But if a is in the set B and in the set C, then it is in the set $B \cap C$. Thus, we have that a is in the set A and $B \cap C$, which means that a is in the set $A \cap (B \cap C)$. We can now write all of this in our new notation.

$$a \in (A \cap B) \cap C \Rightarrow a \in A \cap B \text{ and } a \in C$$
$$\Rightarrow a \in A \text{ and } a \in B \text{ and } a \in C$$
$$\Rightarrow a \in A \text{ and } a \in B \cap C$$
$$\Rightarrow a \in A \cap (B \cap C).$$

We have therefore established that $(A \cap B) \cap C \subseteq A \cap (B \cap C)$, since every element of the former must also be an element of the latter, and we note that this is identical to our proof that $(A \cup B) \cup C \subseteq A \cup (B \cup C)$ only with the word "and" replacing every appearance of the word "or."[7] In the exact same way, we can prove that $A \cap (B \cap C) \subseteq (A \cap B) \cap C$. Namely, we have the following logical chain:

$$a \in A \cap (B \cap C) \Rightarrow a \in A \text{ and } a \in B \cap C$$
$$\Rightarrow a \in A \text{ and } a \in B \text{ and } a \in C$$
$$\Rightarrow a \in A \cap B \text{ and } a \in C$$
$$\Rightarrow a \in (A \cap B) \cap C.$$

We therefore have that the two sets $(A \cap B) \cap C$ and $A \cap (B \cap C)$ are mutually subsets of each other, and therefore they must be equal as sets.

---

[7]This is not an accident, but rather a reflection of the very close logical relationship between the meaning of the word "and" and the meaning of the word "or." We will have more to say about the generality of these results in a later volume when we use logic to explore logic itself, but for now we just note this fact.

Let us look at an explicit example of this by again taking our sets to be $A = \{1, 2, 3, 4\}$, $B = \{2, 4, 6, 8\}$, and $C = \{2, 6, 7, 8\}$, just as we did at the end of the previous section. We see that $A \cap B = \{2, 4\}$ and therefore that $(A \cap B) \cap C = \{2\}$, since the only element that is in both $A \cap B = \{2, 4\}$ and $C = \{2, 6, 7, 8\}$ is the element 2. Similarly, we see that $B \cap C = \{2, 6, 8\}$ and therefore that $A \cap (B \cap C) = \{2\}$, since the only element that is in both $A = \{1, 2, 3, 4\}$ and $B \cap C = \{2, 6, 8\}$ is the element 2. We therefore have, in this particular example, that the two sets $(A \cap B) \cap C$ and $A \cap (B \cap C)$ are equal, as they should be.

Let us combine the main result of this section with the main result of the previous section to state the following theorem.

Theorem 17.1. *Let A, B, and C be three sets (any sets at all). Then*

$$(A \cup B) \cup C = A \cup (B \cup C)$$

and

$$(A \cap B) \cap C = A \cap (B \cap C).$$

One of the consequences of this theorem is that we can write the union or intersection of any three sets without any parentheses at all, because we know that no matter which two we "union first" or "intersect first," we will always get the same result once we include the third set. We can therefore write $A \cup B \cup C$ and $A \cap B \cap C$ without any ambiguity as to the logical content of these symbolic expressions **only because** we have already shown that it does not matter which two sets we combine first. This is a very important bit of logic to understand. We did not **initially** define what an expression like "$A \cup B \cup C$" means because we never defined what the union of three sets is — we only defined what a union of two sets is. We then **proved** that if we form a union of three sets by taking the union of any two of them first, and then taking the union with the third, then the resulting set does not depend on which two we chose to take the union of first. **Only then** were we allowed to write an expression like $A \cup B \cup C$ and have it be a well-defined expression.

In the previous section we saw that unions "play well with themselves" and in this section we saw that intersections also "play well with themselves." In the next section we will turn our attention towards examining how unions and intersections play with **each other**, and in doing so we will develop the tools that we need for manipulating arbitrary combinations of unions and intersections of arbitrarily many sets.

Putting It All Together — 3 Sets

Two sections ago we asked what would happen if we took the union of two sets A and B and then took another union of this set with a third set C, and in the previous section we asked what would happen if we took the intersection of two sets A and B and then took another intersection of this set with a third set C. Let us now ask what would happen if we took the union of two sets A and B and then took the **intersection** of this set with a third set C. Namely, suppose we first form $A \cup B$ out of two sets A and B and then we took the intersection of this set with C to form the set $(A \cup B) \cap C$. Our goal is to re-express $(A \cup B) \cap C$ in terms of other combinations of unions and intersections, just as we were able to re-express $(A \cup B) \cup C$ as $A \cup (B \cup C)$. Our motivation for this is two-fold. First, having different expressions of the same object is always useful, since in any given problem or proof one expression might be more useful and/or enlightening than another expression. And second, it is interesting. That is enough motivation in itself.

Let us therefore ask ourselves what kind of elements are in $(A \cup B) \cap C$. From the definition of the intersection of two sets, if an element a is in the set $(A \cup B) \cap C$ then a must be in the set $A \cup B$ **and** in the set C. But if a is in the set $A \cup B$, then from the definition of the union of two sets, a must be in A **or** in the set B. Thus we have found that a is in A or B, and **either way** it is also in C. This means that either a is in A and C, or that a is in B and C. Now if a is in A and C, then a is in $A \cap C$, and if a is in B and C, then a is in $B \cap C$. Thus, from the sentence preceding the last sentence, we have that a is in $A \cap C$ or that a is in $B \cap C$. This, however, is the very definition of having a be in the set $(A \cap C) \cup (B \cap C)$. We have therefore shown that if a is in the set $(A \cup B) \cap C$, then it must also be in the set $(A \cap C) \cup (B \cap C)$. In words, this says if a is in the intersection of "the union of A and B" with C, then it is also in the union of "the intersection of A and C" with "the intersection of B and C." In symbols, this means that we have proven that

$$(A \cup B) \cap C \subseteq (A \cap C) \cup (B \cap C).$$

Before proving that these two sets are actually equal, let us summarize the argument of the previous paragraph using our fancy new streamlined notation. We have that[8]

$$\begin{aligned}
a \in (A \cup B) \cap C &\Rightarrow a \in A \cup B \text{ and } a \in C \\
&\Rightarrow a \in A \text{ or } a \in B, \text{ and } a \in C \\
&\Rightarrow a \in A \text{ and } a \in C, \text{ or } a \in B \text{ and } a \in C \\
&\Rightarrow a \in A \cap C \text{ or } a \in B \cap C \\
&\Rightarrow a \in (A \cap C) \cup (B \cap C).
\end{aligned}$$

We now want to show that the two sets $(A \cup B) \cap C$ and $(A \cap C) \cup (B \cap C)$ are actually equal, and in order to do this all that remains to be shown is that the latter is a subset of the former (since we have already shown that the former is a subset of the latter). In particular, we need to show that any element that is in $(A \cap C) \cup (B \cap C)$ is also in $(A \cup B) \cap C$. The logic for this is precisely the same as that which is given above, only now in the reverse direction. Namely, we have that

$$\begin{aligned}
a \in (A \cap C) \cup (B \cap C) &\Rightarrow a \in A \cap C \text{ or } a \in B \cap C \\
&\Rightarrow a \in A \text{ and } a \in C, \text{ or } a \in B \text{ and } a \in C \\
&\Rightarrow a \in A \text{ or } a \in B, \text{ and } a \in C \\
&\Rightarrow a \in A \cup B \text{ and } a \in C \\
&\Rightarrow a \in (A \cup B) \cap C.
\end{aligned}$$

This therefore proves that $(A \cap C) \cup (B \cap C) \subseteq (A \cup B) \cap C$, and therefore when we combine this with our result above we see that we have successfully proven the following theorem.

---

[8] Note that the comma in the second line of the following argument is extremely important. It is meant to clarify that a is in A or a is in B, and that **either way** a is in C.

Theorem 17.2. *Let A, B, and C be three sets. Then*

$$(A \cup B) \cap C = (A \cap C) \cup (B \cap C).$$

Let us recap what we have done so far in this chapter. We first saw what happens when we take a union of two sets and then take another union of the resulting set with a third set. We then saw what happens when we take an intersection of two sets and then take another intersection of the resulting set with a third set. Finally, we just finished establishing what happens when we first take a union of two sets and then take an **intersection** of the resulting set with a third set. The only thing that remains to be established is what happens when we first take an intersection of two sets and then take a union of the resulting set with a third set. Namely, we must still explore the set $(A \cap B) \cup C$. We will explore this set in the first exercise at the end of this chapter, where we will prove the validity of the following theorem.

Theorem 17.3. *Let A, B, and C, be three sets. Then*

$$(A \cap B) \cup C = (A \cup C) \cap (B \cup C).$$

We will also begin to see in the exercises that Theorems 17.2 and 17.3 are all we need to be able to simplify any combination of an arbitrary number of unions and intersections. We will begin to get the sense that, in a very precise sense, the operations of union and intersection behave amongst sets just like the operations of addition and multiplication behave amongst numbers. Indeed, we will see later in this series that we can make this statement completely precise, but for now we will just reflect on how surprising this fact is. Namely, that operations as abstract as "union-ing" and "intersecting" arbitrary sets can behave similarly to operations as concrete and familiar as adding and multiplying numbers. To see this explicitly we must do the exercises at the end of this chapter, so let us turn to these now. In the next couple of chapters we will go on to see more ways of taking two sets and making a third set from them. These constructions of new sets from old ones will differ greatly from the constructions of unions and intersections, and will be extremely important for us in the future. We can think of these few chapters as building the machinery that we will need to explore increasingly complex and intricate mathematical structures — namely, we our increasing the size of our mathematical toolbox so that we can go on to build cooler and cooler things!

Exercises

1) Prove Theorem 17.3.

Note: The following two exercises are the hardest in this whole volume. To truly understand them it will likely require a serious time commitment as well as a decent amount of pen and paper. They are both important for gaining the elusive quality of "mathematical sophistication," though they are not absolutely necessary for the remainder of this volume. Namely, none of the structures or ideas that we will introduce in the remainder of this volume will rely on the results of the following two exercises. That said, these two exercises provide good practice for manipulating sets, and these skills will indeed be necessary for understanding much of the math that we will explore in future volumes. Therefore, we recommend that these exercises are understood **eventually**, but a deep and thorough understanding of them is in no way required at this moment. Thus, if we desire to more casually work our way through this volume then these exercises do not need to be poured over.

2) Use Theorem 17.2 to prove that[9]

$$(A \cup B) \cap (C \cup D) = (A \cap C) \cup (A \cap D) \cup (B \cap C) \cup (B \cap D).$$

Hint: We will have to use Theorem 17.2 twice, and in the first instance it is important to view one of the two unions on the left side of the equation as a **single** set when applying the theorem. (Note: This one is hard. Try to gain some intuition for what both sides of the equality mean in words, and that will help guide the symbolic manipulation. Remember, we work this out explicitly in the back of this volume.)

3) Use Theorem 17.3 to prove that

$$(A \cap B) \cup (C \cap D) = (A \cup C) \cap (A \cup D) \cap (B \cup C) \cap (B \cup D).$$

The same hint and note from the previous problem apply to this problem as well.

---

[9] The following expression will likely seem rather daunting at first sight, so we are encouraged to take a deep breathe and remain calm. As we will see in the solution to this problem in the back of this volume, the actual meaning behind all of this is not all that scary. This is a great time to reread or at least think about the "Note On Notation" at the beginning of this volume, and to recall that all of these symbols are only representing ideas that we have already become familiar with, and which we just have to work through slowly. Namely, every parenthesis and every symbol means something — and more importantly we already know what all of these things mean — so we just have to slowly and calmly work our way through the expressions and unpack the meaning from them one step at a time. We do this explicitly in the solutions.

Chapter 18

More Constructions, Part 1

The Cartesian Product

In Chapter 16 we explored two different ways that we can "build" new sets from old ones. Namely, we defined what it means to take the union and the intersection of two different sets. We then went on to explore how these operations interact with each other and developed a sort of "arithmetic of sets." We are now going to return to our initial question in Chapter 16 and see whether or not there other ways that we can build new sets from old ones. It turns out that there are lots of things that we can do, but it just so happens that there are a few constructions that are particularly useful and important. Unions and intersections are two of the most fundamental constructions in all of set theory, and in the current and next chapters we will introduce two more — the Cartesian product and the disjoint union. Both of the latter two constructions, however, are different from the constructions of unions and intersections in important ways. Let us first see what these differences are.

Recall that when forming the union or intersection of two sets we never altered the individual elements of the respective sets. Namely, we simply constructed a new set from the elements that were in the two old sets, while never making any changes to the actual elements themselves. The union of two sets A and B has as its elements all of the elements that are in either A or B, and the intersection of two sets A and B has as its elements all of the elements that are in **both** A

THE CARTESIAN PRODUCT

and B. In neither construction are we actually changing the elements themselves — we can think of unions and intersections as simply "redistributing" the elements of two sets. In the two constructions that we will study in this and the next chapter (Cartesian products here and disjoint unions in the next), we **will** be altering the elements themselves, but we will be doing so in an extremely natural and obvious way.

We will begin with the Cartesian product of two sets. As usual, let us call our initial two sets A and B. A and/or B can be infinite sets, finite sets, sets of numbers, sets of animals, sets of fruit, sets of any combination of these objects, or sets of anything else at all. Let us further suppose that a is an element in A (recall from Chapter 16 that this is written $a \in A$), and that $b \in B$. We can then define a **new element**, (a, b), that we simply define to be "the pair of elements a and b." This is completely abstract. We can take two concrete objects like a basketball and an apple and form the abstract pair of elements (basketball, apple). This pair of elements is a much more abstract element than either of the individual elements that make up the pair, but this is not at all a problem. There is nothing stopping us from viewing a pair of objects as a single object — indeed, this is somewhat analogous to how we can view entire sets as single elements in order to form sets of sets.

What we do to construct the Cartesian product of two sets A and B is construct all possible pairs of elements where one element in each pair comes from A and the other element comes from B. If $A = \{1, 2, 3\}$ and $B = \{\text{donkey}, \text{chicken}, \text{cow}\}$, then the pair of elements 1 and chicken is simply defined to be (1, chicken), and the pair of elements 2 and donkey is simply defined to be (2, donkey). To form the Cartesian product of A and B, we simply need to include all such pairs in our new set. Let us make this a formal definition before we explore some examples.

Definition 18.1. Let A and B be sets. The Cartesian product of A and B is the set of pairs (a, b) such that $a \in A$ and $b \in B$. We denote the Cartesian product of A and B as $A \times B$, so that we have $A \times B \equiv \{(a, b) \mid a \in A \text{ and } b \in B\}$.

Note that we used the symbol "$\equiv$" in the last line of the above definition because this is how we are defining the set, and this symbol is read "is defined to be," as discussed in Chapter 16. Note also that the string of symbols that end the definition above should be read as "$A \times B$ is defined to be the set of all elements (a, b) such that a is an element of A and b is an element of B." It is always important to make sure our math is grammatically correct, and to remember how our symbols line up with the logical ideas that they represent. Namely, the "$\equiv$" symbol gives us the phrase "is defined to be," the brackets "$\{\}$" give us the phrase "the set of all elements," and the vertical line "$|$" gives us the phrase "such that."

This definition will make more sense if we see an example. As usual, let us take a simple example first. Let us let $A = \{1, 2, 3\}$ and $B = \{a, b, c\}$. Then

$$A \times B = \{(1, a), (1, b), (1, c), (2, a), (2, b), (2, c), (3, a), (3, b), (3, c)\}.$$

This is because these 9 elements are all the possible pairs of elements that we can make once we require that the "first slot" in each pair is some element in A, and that the "second slot" in each pair is some element in B. It is also not an accident that there are 9 elements in $A \times B$, where 9 happens to be the number of elements in A times the number of elements in B (namely, $3 \times 3 = 9$). The reason this is not an accident is that in order to form all possible pairs of elements where one element comes from A and one comes from B, we must do the following. We must take each element of A and form a pair for each element of B. Namely, if we take $1 \in A$, then we need to form the pairs $(1, a)$, $(1, b)$, and $(1, c)$. These are all of the possible pairs that have the element 1 in them. We then go on to another element in A (for example, 2) and do the same. Thus for each element in A we will get three pairs in total, and since there are three elements in A we will have $3 + 3 + 3 = 3 \times 3 = 9$ total pairs in the set $A \times B$.

Indeed, from this example we can see that if A and B are both finite sets, and if A has N elements and B has M elements, then the Cartesian product $A \times B$ of A and B has NM elements.[1] This can be seen by using the same logic as in the previous paragraph. Namely,

---

[1] Recall that when we just stick two numbers next to each other like "NM," this is just an abbreviation for multiplying N by M.

we see that for every element in A, we get M elements in $A \times B$. This is because if $a \in A$ then there are exactly M elements of the form $(a, \cdot)$ where the "$\cdot$" cycles through the M elements in B. Since there are N elements in A, and since each of these N elements corresponds to M elements in $A \times B$, we see that $A \times B$ has NM elements. To be completely thorough, we can see this result by noting that if A has N elements then we can write $A = \{a_1, a_2, ..., a_N\}$ where each a_i corresponds to some element in A (these could be numbers, or animals, or NBA players, or any combination thereof). Similarly, if B has M elements, then we can write $B = \{b_1, b_2, ..., b_M\}$. We then have that any pair in the set $A \times B$ can be written as (a_i, b_j) where i is any number between 1 and N, and where j is any number between 1 and M. This means that for **each** of our N choices of i, we get M choices for j, meaning that we get NM choices total. Since each choice corresponds to a different element of $A \times B$, we have again confirmed that the set $A \times B$ has NM elements in it whenever A is finite with N elements and B is finite with M elements. Let us make this a formal proposition so that we can refer back to it quickly if the need arises.

Proposition 18.2. Let A be a finite set with N elements, and let B be a finite set with M elements. Then $A \times B$ has NM elements.

We should not let this proposition fool us into thinking that we can **only** take Cartesian products of finite sets, because this is most certainly not the case. We can easily take the Cartesian product of two infinite sets (or of an infinite set with a finite set). This would only mean that we would have infinitely many elements in the resulting Cartesian product, but there is nothing wrong with that.

A Look Towards The Future

There are some important things to note about the construction of a Cartesian product, which will be important for us **much** farther down the road when we introduce and study categories.[2] The

---

[2] A category is a particular, extremely gorgeous, and wildly abstract mathematical structure. We will not take a detailed look at them for several volumes,

first thing to note is that the elements in the Cartesian product are completely abstract — even more abstract than the elements of the original sets that we used to form the Cartesian product. One good way to think of the Cartesian product of two sets is to consider the product as "the set of **ways** to pick one element from one set and one element from the other." This is a very abstract kind of element because it is not a "thing" like an apple or an orange or a number, but rather a "way" of choosing elements. We must view a "way of choosing two elements" as a single element. This perspective on elements of a Cartesian product will make the next few paragraphs more digestible as well.

The next thing to note is that there is a certain amount of arbitrariness to our definition of the Cartesian product of two sets. Namely, why did we define the Cartesian product of A and B to be the set of elements written as (a, b) (with $a \in A$ and $b \in B$) as opposed to the set of elements written as $a \times b$? Or why not the set of elements written as $[\{< a >\}\{< b >\}]$ with $a \in A$ and $b \in B$? Or why not reverse the order of the elements in a given pair so that the elements of B are on the left and those of A are on the right?

The answer to all of these questions is that we very well **could** have defined the Cartesian product of A and B in any of these other ways. However, there is in a very clear sense the notion that all of these sets would somehow be "the same." Namely, these different possible definitions of $A \times B$ all contain the same information, and we are just finding different ways to write that information down. In particular, the set of all pairs (a, b) with $a \in A$ and $b \in B$ is in a very clear sense "equivalent" to the set of all pairs $[\{\{[a, b]\}\}]$ with $a \in A$ and $b \in B$ — the only difference between these two sets would be how we write them down.

Clearly, mathematics should not be dependent on how we choose to write things down, so how do we reconcile all of this? In short, we will not reconcile this. At least not right now. In order to properly reconcile this we will need to learn about categories, but as mentioned above we will not explore categories for quite a while. The important

---

but the concepts that we will learn between now and then are essential for gaining a deep appreciation for why categories are interesting. For now, we simply must be patient.

thing to keep in mind is that while all of these alternative ways of writing the elements in a Cartesian product are possible, we are simply **defining** one of them to actually be the Cartesian product. We then simply remember that if the enemy hands us one of these other sets (like the set of all pairs $[\{\{[a,b]\}\}]$ with $a \in A$ and $b \in B$) then we can very naturally identify it with the Cartesian product that we have defined.

To summarize, the Cartesian product of two sets A and B is the set obtained by taking all possible **pairs** of elements, one from A and one from B. There are several ways to **write** these pairs down, but at the end of the day the **mathematical content** in each of these various ways is identical. Let us now take a look at the following exercises in order to gain some more familiarity with Cartesian[3] products.

Exercises

1) Let $A = \{1, 2\}$, $B = \{a, b\}$, and $C = \{3, 4\}$. Then the Cartesian product $A \times B$ of A and B is a perfectly good set, and so we can take its Cartesian product with C, denoted by $(A \times B) \times C$. Write down one example of an element in the set $(A \times B) \times C$. How many elements are in $(A \times B) \times C$?.

2) Let $A = \{1, 2, 3\}$. Write down all of the elements in $A \times A$. How many elements are in $A \times A$?

3) Let $A = \{1, 2\}$, $B = \{a, b\}$, and $C = \{b, c, d\}$. Write down all of the elements in the set $A \times (B \cup C)$. How many should there be? Also write down all of the elements in the set $A \times (B \cap C)$. How many should there be in this case?

---

[3] As a side note, we call this the "Cartesian" product because it is named after the mathematician/philosopher Rene Descartes, and because great mathematicians get mathematical structures and/or theorems named after them.

Chapter 19

More Constructions, Part 2

The Disjoint Union

So far we have studied three ways of creating new sets from old ones. Namely, given two sets A and B, we can consider their union $A \cup B$, their intersection $A \cap B$, and their Cartesian product $A \times B$. We are now going to study a fourth way of creating a new set from two old ones. The construction that we will introduce in this chapter is going to be most similar to the union that we have studied before. In fact, this construction will be so similar that we even keep the word "union" in the name of the construction — we will call this construction the **disjoint union** of two sets A and B. Despite the fact that the disjoint union (which we have yet to define) is very similar to the union, it will of course not be exactly the same (otherwise we would not be spending a separate chapter studying it).

The ways in which the disjoint union is different from the usual union might seem trivial at first, but these differences will actually be very important for us farther down the road. Indeed, just as the construction of the Cartesian product will end up being extremely important for us in the future when we study categories, so too will the construction of the disjoint union. Unfortunately we will not be able to see precisely why this is the case until we properly study category theory in a much later volume, but this construction will still have utility for us in the nearer future so it is still important for us to introduce it now. Let us begin by first recalling exactly what it

THE DISJOINT UNION

is that the normal union of two sets did for us.

When we took the union of, say, two sets A and B, we formed a set $A \cup B$ by simply "joining these sets together." By this we mean that for the set $A \cup B$, we simply consider the elements of A and the elements of B as being all together. In particular, we do not change the elements of A and B when we form $A \cup B$, we simply form a set by including all of the elements in A and all of the elements in B. But since we do not change the elements when we form $A \cup B$, any repeated elements (i.e., any elements that are both in A and in B) are simply considered as one element in the union. For example, the union of $\{1, 2, 3\}$ and $\{4, 5, 6\}$ is $\{1, 2, 3, 4, 5, 6\}$, while the union of $\{1, 2, 3\}$ and $\{1, 2, 4\}$ is $\{1, 2, 3, 4\}$. The important thing to note here is that in the second example the elements 1 and 2 are in **both** of the initial sets, yet they only make one appearance in the union. This is because — from the very definition of a set — we do not distinguish between identical elements in any given set.[1]

This sets us up nicely to make the distinction between a union and a disjoint union. In short, the disjoint union of two sets will be like a union that "remembers" which set each element comes from. Indeed, we will "label" our elements before union-ing them, so that elements that appear in both initial sets will still be distinguished in the resulting set. One result of this construction will be that if we take the disjoint union of a finite set having N elements with a finite set having M elements, then the resulting set will have $N + M$ elements. As we saw in the preceding paragraph, this statement is **not** true about the normal union. In particular, the union of the sets $\{1, 2, 3\}$ (which has 3 elements) and $\{4, 5, 6\}$ (which also has 3 elements) does indeed have 6 elements (which is the sum of the number of elements of each constituent set), whereas the union of $\{1, 2, 3\}$ and $\{1, 2, 4\}$ only has 4 elements (which is **not** the sum of the number of elements of each constituent set).

Let us look at a simple example of a disjoint union before making

---

[1] One corollary to this is that the union of any set with **itself** is simply itself again. We noted this in the solution to the second problem of the previous chapter, so let us now just take a look at an example. Consider the union of $\{1, 2, 3\}$ and $\{1, 2, 3\}$. The set whose elements are in either of these two sets is simply $\{1, 2, 3\}$ again, so that we have $\{1, 2, 3\} \cup \{1, 2, 3\} = \{1, 2, 3\}$!

a formal definition of this construction. Let $A = \{1\}$ and $B = \{1\}$. We then have that $A = B$, and therefore that $A \cup B = \{1\}$ as well. Let us now label these elements differently. Namely, let us label the 1 that comes from A as 1_A, and let us label the 1 that comes from B as 1_B. It is important to remember that this is only a labeling — it is just a new notation for the abstract idea of 1 that we had originally. Namely, the abstract idea is not changing, it is just the notation that is changing, and that is completely fine. Then the disjoint union of A and B will be the set whose elements are $\{1_A, 1_B\}$. We should view this set as having two different "copies" of the element 1, in such a way that we can "remember" which set each 1 comes from.

Other than its future application to category theory, the disjoint union actually has some utility for us immediately. Namely, this construction allows us to deal with sets in a much more flexible way than the original definition of a set allows. Namely, we have found a logical way around the requirement that sets cannot distinguish between identical elements.[2] For example, we would typically have that $\{1, 1, 2, 3\} = \{1, 2, 3\}$ since sets do not "see" the duplicate elements. However, motivated by the example given in the previous paragraph, we can construct a set using disjoint unions that might look something like $\{1_A, 1_B, 2, 3\}$, and this would **not** equal $\{1, 2, 3\}$ because we are now "remembering" where the duplicate 1's come from.

Now that we have seen a little motivation for why this definition is useful, let us finally make the formal definition of the disjoint union of two sets. In the following definition we will not label our elements using subscripts. Instead, we will label which set a given element comes from using the notation (a, A) (instead of a_A) whenever a is an element of the set A.

Definition 19.1. Let A and B be sets. The disjoint union of A and B is the set, often denoted by $A \sqcup B$, of elements of the form (a, A) or (b, B) where $a \in A$ and $b \in B$. We thus have, in symbols,[3]

$$A \sqcup B = \{(a, A) \mid a \in A\} \cup \{(b, B) \mid b \in B\}.$$

---

[2] And we have done so without going against any of the logic that we have already established.

[3] Remember, the following is not so scary, we just need to slowly work our way through it one symbol at a time.

All that this definition means is that the disjoint union $A \sqcup B$ of two sets A and B is constructed by first forming the elements (a, A) and (b, B) where $a \in A$ and $b \in B$ — which is nothing but labelling the elements in each respective set — and then taking the **normal** union of all of these elements. For example, let us consider the case when $A = \{1, 2, 3\}$ and $B = \{2, 3, 4\}$. In order to form the disjoint union $A \sqcup B$, we must first form the elements $(1, A)$, $(2, A)$, $(3, A)$, $(2, B)$, $(3, B)$, and $(4, B)$. We then just take the normal union of all of these elements so that we have

$$A \sqcup B = \{(1, A), (2, A), (3, A), (2, B), (3, B), (4, B)\}.$$

We note that it is indeed the case that $A \sqcup B$ has 6 elements, whereas $A \cup B$ only has 4 elements. This is because the sets A and B both have the elements 2 and 3, and so in the **normal** union these elements do not get distinguished whereas in the **disjoint** union we do count one copy of 2 from the set A and one copy of 2 from the set B (and the same goes for the element 3 as well).

We therefore see that the disjoint union of two sets is the set of all elements from each set, where we also remember where each element "comes from." In some sense, then, this is actually the **more** natural notion of a union of two sets than our original definition of union, since the disjoint union preserves all of the information of the two original sets. However, whichever notion of union one thinks is "more natural" is purely a matter of taste, and regardless of one's taste it is important to understand both notions of union, as both are completely fundamental in mathematics. Let us therefore go on to explore one subtlety involved with the construction of a disjoint union that has been hiding from us all along.

An Important Subtlety

One question that remains to be asked is what happens if we want to take the disjoint union of a set with itself? In particular, we can indeed construct the union, intersection, and Cartesian product of any set A with itself. This is because the union of A with A is "the set of all elements that are in A or in A" and therefore $A \cup A = A$.

Similarly, the intersection of A with A is "the set of all elements that are in both A and A," and therefore $A \cap A = A$. Finally, the Cartesian product of A with A is the set of all elements (a, b) such that $a \in A$ and $b \in A$. So now we should ask if and how we can form the disjoint union of A with itself. The reason that this is a subtle question is that the disjoint union of two sets relies on there being some kind of "labeling system" that allows us to "remember" where all of the elements "come from." When the sets were A and B this was easy because we could just label the elements with A and B, respectively.

So what do we do if we want to construct the disjoint union of A with itself? Technically speaking, we cannot quite do this just yet. For if we have two "copies" of A and we label all of the elements as (a, A), then we still have no way of distinguishing which "copy" of A the elements come from. One important quality that we need to recover is that the disjoint union of A with itself needs to have twice as many elements as A (if A is a finite set, of course), and if we label the elements a in A simply as (a, A) then the disjoint union of A with itself would have the same number of elements as A. What is missing therefore is a labeling system that distinguishes which **copy** of A the elements are coming from. Accordingly, we can simply add in a "labeling set" to distinguish the different copies of A.

For example, if $A = \{a, b, c\}$, then we can simply rename the set and form two copies of it: $A_1 = \{a, b, c\}$ and $A_2 = \{a, b, c\}$. Then we can write the elements in the disjoint union of A with itself[4] as, for example, (a, A_1) and (a, A_2). In other words, we are simply **forcing** ourselves to remember which set these elements come from by introducing this new labeling of the sets.

In the above example our "labeling set" was just $\{1, 2\}$, since we chose the subscripts 1 and 2 as a way of distinguishing the two copies of A. However, this "labelling set" could have been anything. If our "labeling set" were {Kobe, LeBron}, then we would simply write A_{Kobe} and A_{LeBron}. In this case, if $A = \{1, 2, 3\}$ and we chose this "Kobe and Lebron" labelling system, then we would be able to write

---

[4] Namely, in the set $A \sqcup A$.

the disjoint union of A with itself as

$$A \sqcup A = \{(1, A_{\text{Kobe}}), (2, A_{\text{Kobe}}), (3, A_{\text{Kobe}}),$$
$$(1, A_{\text{LeBron}}), (2, A_{\text{LeBron}}), (3, A_{\text{LeBron}})\}.$$

We note that it is indeed the case that the disjoint union of A with itself has 6 elements in this example, which is twice the number of elements that A has. This is precisely what we wanted.

We can now see that the disjoint union construction is simply the following. If we are taking the disjoint union of two sets A and B, then we simply label each element by its respective set and then take the **normal** union of all of these elements. If we want to take the disjoint union of some set A with **itself**, then we simply choose a way to label two different copies of A, and then label the elements in each copy according to this choice and take the normal union of these elements.

The moral of this story is therefore that we indeed **can** take the disjoint union of sets with themselves, we just have to get a little creative. Namely, we have to change what we call the copies of the original set so that we have a distinctive way of labeling the elements in the disjoint union. If this seems unnatural, then we should try to convince ourselves that it is not. After all, we have already had to change what we call our **elements** when we form Cartesian products and disjoint unions, so why not allow ourselves to change what we call our **sets**? Moreover, all of this "changing what we call stuff" is just notation, and we have seen time and time again that notation is not what matters, but rather the ideas that the notation represents. Thus, once we have the ideas clear, we just need to find a good notation to reflect those ideas, and it just so happens that the idea of taking the disjoint union of a set with itself requires a little more creativity in its notation. The **underlying idea** of a disjoint union is simple though — we just want to remember where each element comes from when we take the union of two sets — and the rest is just formal notation. The exercises at the end of this chapter will help us solidify our understanding of disjoint unions, as well as get us more used to these notational formalities.

A Pause In Construction

This temporarily ends our investigation of constructing new sets from old ones. However, we should not think that the union, intersection, Cartesian product, and disjoint union of two sets are the only possibilities for constructing sets! There are lots of other ways to bring sets together to form new ones, and some of these other ways are extremely important. We will indeed see these other constructions in due course, but for now we will move on to other considerations simply to shake things up.

Namely, we will now move on (in the next chapter) to exploring a very particular application of the various constructions that we have introduced in this and the previous few chapters. In particular, we will be able to use a lot of what we have learned so far to look at ideas from basic high school or middle school math in a whole new light. This new light will be a much more abstract one, and it is precisely this abstraction that will help us clear up what we have "really been doing" all along in middle and high school math class! These parallels to more familiar[5] math are important for us, as they highlight the fact that the kind of math that we are currently exploring has always been and will always be "behind the scenes" of what is more "conventional." Moreover, it is always enlightening to see how certain mathematical explorations are simply "special cases" of a larger, more general framework. The framework that we have introduced thus far in this volume is incredibly flexible, general, and abstract, and the high school math that we will make a connection with in the next chapter is simply a **very particular** example of all this abstract machinery. However, as we have already seen and as we will continue to see, the mathematical world is **much** larger than that which we learned in high school, and chapters like the next one are simply here to help us appreciate this fact.

---

[5]Where by "familiar" we simply mean that which is currently commonly taught in middle and/or high school math classes.

Exercises

1) Let $A = \{1,2\}$, $B = \{2,3\}$, and $C = \{1,3\}$. How many elements are in $(A \sqcup B) \sqcup C$? How many elements are in $A \sqcup (B \sqcup C)$? Are the sets $(A \sqcup B) \sqcup C$ and $A \sqcup (B \sqcup C)$ equal? (This is a very subtle question, and is indeed a warm-up for the mysterious category theory that we have been alluding to for a while. Therefore **solving** this problem is not so important at this stage, but understanding its solution (which is given in the back of this volume) is quite enlightening. Additionally, the solution to this problem can be viewed as a chapter in its own right — both in terms of length as well as importance. Thus, we should be ready to devote a good amount of time to this solution.)

2) Let A, B, and C be as in the first exercise above. How many elements are in the set $(A \cup B) \sqcup C$? Explicitly write down an element (any element) from this set. (Recall that this is the disjoint union of the set $A \cup B$ with the set C.)

3) Again let A, B, and C be as in the first exercise above. How many elements are in the set $(A \cap B) \sqcup C$? Explicitly write down an element (any element) from this set.

4) Yet again let A, B, and C be as in the first exercise above. How many elements are in the set $(A \times B) \sqcup C$? Explicitly write down an element (any element) from this set.

Chapter 20

Reconsidering High School Algebra

The Foundations

A very large part of mathematics involves taking a mathematical structure that we are familiar with and generalizing it to some "larger" structure. By "larger," we simply mean "more general," or equivalently "more abstract." Whenever we make such a generalization, our original structure becomes a special case of this new, larger structure. Accordingly, we want to make sure that we "recover" our original object when we consider the corresponding special case.

For a non-mathematical example, let us again consider the great NBA player Kobe Bryant. We know that Kobe Bryant is[1] an NBA player, so that any statement that we can prove about all NBA players will also be true about Kobe Bryant. However, a little bit of thought will convince us that NBA players are also athletes, and that athletes are more general objects than NBA players simply because any NBA player is an athlete, yet there are lots of athletes that are not NBA players. Thus, any statement that we can prove about athletes in general will also be true about NBA players, and therefore will also be true about Kobe Bryant. Even more generally, any statement that we prove about human beings will also be true about athletes

---

[1] And possibly "was" at the time of our reading this book.

(assuming all athletes are humans), and thus about NBA players and thus about Kobe Bryant.

What we mean by being able to "recover" our original object throughout this process of generalization is that we do not want to lose sight of what we initially wanted to study. For example, if we initially wanted to study facts about Kobe Bryant, it might be fruitful to try to prove things about the more general classes of objects like humans or athletes, because Kobe is a special case — i.e., a **particular instance** — of both a human and an athlete. However, it would **not** be fruitful to try to prove something about antique wooden tables, simply because Kobe Bryant is not a particular instance of an antique wooden table.

This is a very common mode of thinking for mathematicians (though mathematicians usually focus on non-Kobe-related topics). Namely, we take some object that we want to study, and instead of studying this object directly we try to first generalize it to some "larger" object. The benefits of this are two-fold. First, it is (surprisingly) sometimes easier to prove things about the more general objects than it is to do so with the more specific object. For example, it might be easier to prove some statement about atheletes in general, as opposed to Kobe Bryant in particular. The second benefit is that once we prove something about this larger, more general object, we have a much more powerful statement at our disposal simply because the newly acquired truth applies to a larger class of objects. For example, if we begin by studying Kobe and end up being able to prove something about all athletes, then we can take this truth and apply it to Muhammad Ali and Lionel Messi all the same. So far in this volume we have been able to prove things[2] about completely general sets, which means that **all** of these truths will still hold for **any** mathematical structure that we study, so long as that mathematical structure is some kind (any kind) of set. This is the power of abstraction.

It is quite often the case that generalizing our mathematical structures in this way leads to new and deeper insight into the original structures themselves, as we then have a better idea of what is "re-

---

[2]For example, all of the statements about unions, intersections, Cartesian products and disjoint unions, as well as all of our statements about functions.

ally" going on within these structures. This may all seem a bit abstract right now, but in this chapter we will see how we have already generalized a mathematical structure that we are likely used to from standard middle school or high school mathematics. And indeed, we will see that our generalizations have allowed us to understand what is "really" going on when we do all of this (possibly dreaded) "standard" mathematics.

It is extremely important to note, however, that our motivations for making these generalizations is **not** to simply give a new perspective on familiar mathematics. Indeed, it should **not** be thought that these generalizations simply give a more interesting perspective on familiar examples, but rather that the familiar examples give a **less** interesting perspective on these generalizations. Namely, it is almost always the case that the generalizations take on a life of their own, and show us that there is a much larger world of mathematics than we could have ever imagined. Our digression in this chapter back into the more familiar parts of math is **solely** meant to help us appreciate just how powerful and abstract our considerations have been thus far, by seeing just how **limited** the more familiar, or "standard" portions of math are with respect to our formalism of sets.

Indeed, virtually all of mathematics — and certainly all of the structures that we will be studying in these volumes until we eventually study categories — are nothing but special cases of sets. In other words, our structures will be sets with added properties to them, and the deepest we can go down the rabbit hole of asking "but what really is this mathematical structure" is the answer "it is a set." We can think of this as an annoying five-year-old constantly asking "but why?" where all we do is change "why?" to "what is it?". For example, we can take a highly complex mathematical structure, like a twelve-dimensional hyper-Kahler manifold,[3] and ask what this object "really" is. We can say that this object "really" is a manifold with some added structure to it. We can then ask what a manifold "really" is, to which the answer would be a topological space with some added structure to it. Continuing to be the mathematical version of our annoying five-year-old will eventually lead is to the answer that

---

[3] Do not worry, we are not meant to know what this terminology means at all right now — nor will we for quite some time.

our original structure is "really" a set (with some added structure to it).

It is in this sense that the theory of sets forms the very foundations of mathematics[4] and we therefore also see what motivates the title of this volume. In future volumes we will go on to study sets with various kinds of structure added to them, and in the process we will define and prove various remarkable properties of an immensely large class of mathematical objects — objects that we likely had no idea existed prior to this endeavor.

For now, let us return to the question of how all of this abstract machinery applies to more concrete and familiar mathematics that we learned in middle and high school. After all, if we claim that the theory of sets is flexible and powerful enough to serve as the very foundations of an overwhelming amount of mathematics, then we better be able to at least recover the math that we learned in high school. In particular, let us bring back what may or may not be a relatively painful memory from high school mathematics — namely, plotting lines on a plane!

A Fresh Perspective On Planes

There is a decent chance that we have at least the vaguest memories of the following type of problem from middle school or possibly high school.

Problem: Plot the following equations on a plane:

$$y = 5x + 6$$
$$y = \frac{4}{3}x - 9$$
$$y = -2x + 17.$$

Whether we remember how to do this, and whether we hated it, loved it, excelled at it, or failed at it back in school, does not matter here. We will not be interested in the various steps that one takes to

---

[4] At least a **very** large part of mathematics.

solve this problem or the various formulae that we should memorize.[5] The questions that we are going to answer instead, and the questions that might have helped us (if we indeed needed help) back in school, are the following: What are we really even doing here? What is "the plane"? What are these lines? Where is this all coming from?

Now, it is possible that these questions might seem so basic that they are hardly even worth asking.[6] Namely, it might seem clear that the plane "really is" the thing shown in Figure 20.1. And it might also seem clear that a line on the plane is just what is shown in Figure 20.2.

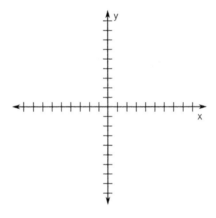

Figure 20.1: The plane.

However, this is not satisfactory. For while pictures are oftentimes very helpful in mathematics, pictures are **not** themselves actual mathematical structures. No proof has ever been completed by only looking at or drawing pictures, and no mathematical statement or structure can be expressed solely as a picture. In math, pictures do not "exist" in any fundamental way — only mathematical structures and logical statements exist. Thus, it seems less than ideal to say that some mathematical structure "is" a picture, or "exists" on a picture. Without pictures though, the question of what we are "really" doing when we are plotting these lines becomes a lot trickier.

---

[5] Perhaps that formula "$y = mx + b$" rings a bell here. If not, that is completely fine, it does not matter.

[6] This might also not be the case, which is entirely okay as well.

A FRESH PERSPECTIVE ON PLANES

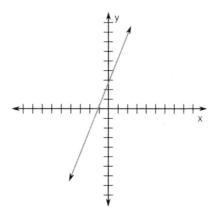

Figure 20.2: A line on the plane.

Luckily for us, however, we already have all of the necessary tools to be able to answer these questions more thoroughly. In this section we will explore what the plane "really is," and in the next section we will explore the same question for the lines that we draw within the plane.

In order to do all of this, we must recall a very important set that we introduced in Chapter 14. Namely, we need to recall that we can define the set of all real numbers, and that this set is often denoted by $\mathbb{R}$. We recall that it is, essentially, the set of all numbers. A bit more precisely, $\mathbb{R}$ is the set of all numbers that can be written as an infinite decimal expansion — this means something like

$$235.35193523562952352353523500030235...,$$

where the "..." simply means that these numbers can go on forever with no rhyme or reason, if we so desire. We note that these include the more standard numbers like "5," where we simply write $5 = 5.0000...$, with 0's repeated forever. We also know that any fraction is also a real number, though this is something that we will just have to take on faith for now.[7] Most importantly, there are **many more** numbers in $\mathbb{R}$ than just the fractions and whole numbers —

---

[7]Recall from Chapter 14 that this fact has indeed been rigorously proven, we just will not discuss the proof in this volume.

indeed, ℝ was the set that we used to find our first example of a "new infinity" in Chapter 14. The set ℝ is referred to as a "continuum" of numbers, because it forms a completely "continuous" set of numbers,[8] allowing us to move from one number to the next "continuously."

Given this "continuous" quality of ℝ, we can relatively convincingly draw this set as a line, just as we did in Chapter 14:

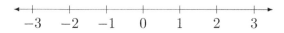

This line goes infinitely far in both directions and has no holes (i.e., it is continuous). This line is simply our **graphical representation** of the set ℝ. The set ℝ itself exists **whether or not** we draw the line — the line just helps our brains to "see" what this set "looks like."[9]

Now we get to use what we have created in this volume. In Chapter 18 we studied the Cartesian product of two sets, which was essentially just the set of all "pairs of elements," one from each set. We also saw that we can take the Cartesian product of a set with itself, thus forming the set of all "pairs of elements in a set." Let us therefore construct the Cartesian product of ℝ with itself. Namely, let us define ℝ × ℝ. We then have a set of pairs of elements, where each individual element in a given pair is an element in ℝ. Namely, we have[10] that

$$\mathbb{R} \times \mathbb{R} = \{(a,b) \mid a, b \in \mathbb{R}\},$$

and in words this reads "ℝ × ℝ is the set of all pairs (a,b) such that both a and b are elements of ℝ."

Now we notice that there is a nice way that we can graphically represent the set ℝ × ℝ. In particular, we saw above that a nice way to graphically represent the set ℝ is via the continuous number line, so now let us take two of these continuous number lines and associate

---

[8]We again place scare quotes around the word "continuous" only because we recall that we have yet to define what we precisely mean by this word. However, it has an intuitive meaning for us and for now this intuitive meaning will have to do. Indeed, we will be able to develop a better notion of "continuity" when we study topology in later volumes.

[9]Scare quotes because a set never "looks like" anything, it is just a set!

[10]Using the notation that we have developed in previous chapters.

A FRESH PERSPECTIVE ON PLANES

one of those lines to one $\mathbb{R}$ that appears in $\mathbb{R} \times \mathbb{R}$, and let us associate the second continuous number line to the second copy of $\mathbb{R}$ in $\mathbb{R} \times \mathbb{R}$. This is shown in Figure 20.3

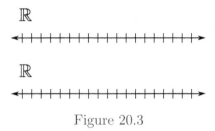

Figure 20.3

We now note that any element (a, b) in the set $\mathbb{R} \times \mathbb{R}$ corresponds to one point on one of these number lines — corresponding to the real number a — and one point on the other number line — corresponding to the real number b. This can be depicted as in Figure 20.4.

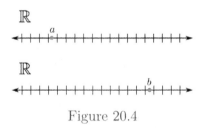

Figure 20.4

However, we must also recall that (a, b) is a **single element** in the set $\mathbb{R} \times \mathbb{R}$, and in Figure 20.4 this single element is being graphically depicted by two separate dots. This is problematic, since it would be nice if our graphical representation could depict the **single** element (a, b) by a **single** point. We can indeed make our graphical representation do this, we just have to get a little creative. Namely, if we tilt one of these copies of $\mathbb{R}$ — let us say it is the "second" copy, referring to the elements in the "second slot" of the pairs (a, b) — so that it makes a 90 degree angle relative to the first copy, then we get a picture precisely like that in Figure 20.1. Namely, what we were calling "the plane" before is now seen to be the graphical representation of the set $\mathbb{R} \times \mathbb{R}$.

What this tilting procedure allows us to do is use the two numbers (a, b) as **coordinates** specifiyng where the **single** point (a, b) lies in the plane. Namely, since we agreed that we will be tilting the "second copy" of $\mathbb{R}$ (so that the number line on which the element b is lying in Figure 20.4 is now vertical on the page), we have that any given pair (a, b) in the set $\mathbb{R} \times \mathbb{R}$ can be represented by placing a **single point** at the location that lies at a in the horizontal direction and b in the vertical direction. An example of this is shown in Figure 20.5.

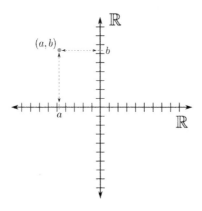

Figure 20.5

Namely, in order to represent the element (a, b) as a point in Figure 20.5, we begin at the point at which the two copies of $\mathbb{R}$ intersect (namely, at the crossing in the figure), and we move to a in the horizontal direction and then to b in the vertical direction. Let us agree that if a is a negative real number then we move to the left, and if a is positive then we move to the right. Similarly, let us agree that if b is a negative real number then we move vertical downwards and if b is positive then we move vertically upwards.[11] For example, the element $(5, -4)$ would be represented by a point that is 5 ticks to the right and 4 ticks downward from the crossing, the element $(0, 7.5)$ would be represented by a point **on the vertical line** (because we must move a distance of zero along the horizontal line) a distance of 7.5 ticks upward from the crossing, and the point of the crossing **itself** corresponds to the element $(0, 0)$ since this element is represented by a point that is a distance of zero away (in any direction) from the crossing, and is therefore the point that is right at the crossing itself.

---

[11]Thus, in Figure 20.5, we know that a is negative and b is positive.

We now note that we have given a precise set-theoretic description of what the plane "really is" — it is a visual **representation** of the precise logical construction of the Cartesian product of $\mathbb{R}$ with itself. Namely, the set $\mathbb{R} \times \mathbb{R}$ is a perfectly good and completely precise logical construction, and the plane that we draw is a graphical representation of this set, where each element in the set (of which there are infinitely many) is represented by a point somewhere on the plane. We have therefore been able to use our abstract machinery to give a thorough answer to the question of what the plane "really is." Let us now turn our attention to doing the same for lines within the plane.

A Fresh Perspective On Lines

At the beginning of the previous section we considered the question of plotting certain equations on the plane. We now have a precise understanding of what the plane "really is" — namely, the plane is a particular graphical representation of the particular set $\mathbb{R} \times \mathbb{R}$ — so let us now consider the equations themselves. For example, we want to know what the equation $y = 5x + 6$ "really" is telling us. It turns out that we can use the machinery that we have already developed to fully answer this question as well.

Namely, what we are "really" doing when we are plotting the line given by the equation $y = 5x + 6$ is specifying a **particular subset** of $\mathbb{R} \times \mathbb{R}$. In particular, this equation is giving a particular relationship between the first and second slots of the pairs (a, b) in the set $\mathbb{R} \times \mathbb{R}$. We will see this by considering an easier example, but before doing this let us change our notation slightly — let us label our pairs by (x, y) instead[12] of by (a, b), simply because this is how it is more commonly presented in textbooks.[13]

We already know that $\mathbb{R} \times \mathbb{R}$ is the set of all pairs (x, y) where

---

[12] It is important to note that this is purely a change of notation, and not at all a change of any logical content.

[13] As a side note, this is also why we usually label our axes by x and y, as in Figures 20.1 and 20.2. Indeed, we very well could have chosen to label our axes by a and b, but for whatever reasons the mathematical community has agreed on using x and y instead (and accordingly, so will we).

both x and y are real numbers. Thus, by putting limitations on the numbers x and y, we can define subsets of the set $\mathbb{R} \times \mathbb{R}$. Some subsets are more interesting than others, however. For example, the subset $\{(0, 4), (1, 1.5)\}$ is a perfectly good subset of $\mathbb{R} \times \mathbb{R}$, but it is just two points in the plane. This is shown in Figure 20.6.

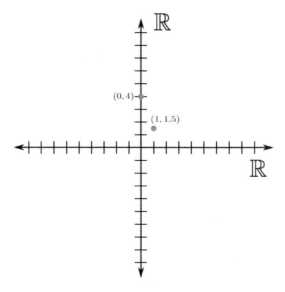

Figure 20.6

However, we could also consider a more interesting subset like "the set of all pairs (x, y) such that y is twice as big as x." If we were to consider this subset then we would see that the pair $(1, 2)$ is in it, because the number in the second slot is indeed twice as big as the number in the first slot. Similarly, $(2, 4)$ is in this subset, as is $(3, 6)$ and $(4, 8)$. Indeed, we quickly see that there are infinitely many elements in this subset. In Figure 20.7 we have shown a few of the points in this subset.[14]

---

[14] To make sense of this figure it is important to note that $(-1, -2)$, $(-2, -4)$, $(-3, -6)$, and so on, are also in this subset because they also share the property that the number in the second slot is two times that in the first slot. Similarly, the pair $(0, 0)$ is in this subset as well because indeed 0 is twice as big as 0. We then must also recall that we have chosen the horizontal line to correspond to the "first slot" and the vertical line to correspond to the "second slot," so that for

A FRESH PERSPECTIVE ON LINES

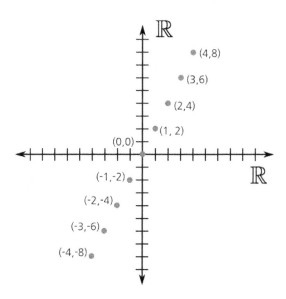

Figure 20.7

The graphical representation of some of the elements in "the set of all pairs (x, y) such that y is twice as big as x" given by Figure 20.7 clearly hints towards the fact that this subset "looks like" a line. And indeed, once we note that all of the points in the plane that lie along the line connecting the dots in the figure are **also** in this subset,[15] then we see that indeed the line drawn in Figure 20.8 corresponds precisely to "the set of all pairs (x, y) such that y is twice as big as x."

We now note that the condition that y is twice as big as x can be more succinctly written simply as $y = 2x$. Namely, the subset that is depicted in Figure 20.8 is "the set of all pairs (x, y) such that $y = 2x$." Moreover, with the notation that we have developed in previous chapters,[16] we note that we can write this subset even more

---

example the element $(1, 2)$ corresponds to the dot that is placed horizontally (to the right) one tick and vertically (upwards) two ticks. As a non-example, we note that the pair $(1, 3)$ is **not** in this subset.

[15] For example, $(1.5, 3)$ is also in this subset and it accordingly lies on the line connecting the dots in the figure. In particular, $(1.5, 3)$ lies on the part of the line connecting the point at $(1, 2)$ to the point at $(2, 4)$.

[16] See the index of notation at the end of this volume for locations of where our

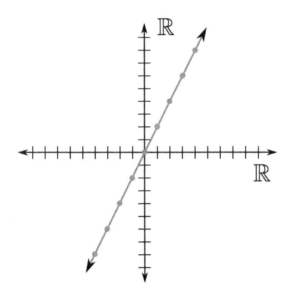

Figure 20.8

succinctly. Namely, if we call this subset S, then we have that

$$S = \{(x, y) \in \mathbb{R} \times \mathbb{R} \mid y = 2x\},$$

which is word-for-word given by the statement that "S is the set of pairs (x, y) in the set $\mathbb{R} \times \mathbb{R}$ such that $y = 2x$." We have therefore given the line depicted in Figure 20.8 a completely precise meaning — it is the visual representation of the subset S of $\mathbb{R} \times \mathbb{R}$ which is defined by the quality that, for each element (x, y) in the subset, the numbers x and y satisfy the requirement that $x = 2y$.

Indeed it happens to be the case[17] that **all lines** in the plane are nothing but visual representations of a particular subset defined by a similar equation. Namely, if the enemy handed us the equation $y = mx + b$ where m and b are any numbers at all,[18] then we can

---

notation was first introduced.

[17] For reasons that we will not explore here. Namely, all of the details of this paragraph are what we will worry about in high school math class. Here, we simply want to understand the abstract and precise mathematical underpinnings to all of this.

[18] For example, in the equation $y = 5x + 6$ we have that $m = 5$ and $b = 6$, and in the equation $y = -\frac{4}{3}x - 9$ we have that $m = -\frac{4}{3}$ and $b = -9$.

A FRESH PERSPECTIVE ON LINES 211

form the subset that is defined to be the set of all pairs (x, y) such that $y = mx + b$. Depending on what m and b are, these lines will look slightly different. For example, it turns out[19] that m has to do with how "steep" the line is. Regardless, what is important here is that all we are doing when we "plot the line given by the equation $y = 5x + 6$" is drawing the graphical representation of the subset of $\mathbb{R} \times \mathbb{R}$ that is defined to have its elements (x, y) satisfy this particular requirement.

This now completely answers the question of what we are "really" doing when we are plotting a line given by some equation — we are simply drawing the graphical representation of a **particular subset** of the plane $\mathbb{R} \times \mathbb{R}$ which is given by this equation. The **reason** why we care about such subsets is not important to us here. Instead, we only care about the precise definition of what we are "really" doing, and now we know.

We should also point out that there are lots of other perspectives from which we can view the issue of plotting lines on a plane — we have simply explored one so far. Another common perspective is to view all of this as the study of certain functions from $\mathbb{R}$ to $\mathbb{R}$. Our drawings then visually represent these functions by showing the entire domain and codomain (both of which are $\mathbb{R}$ here) of the function simultaneously, and drawing a mark on the points corresponding to each other via this function. This perspective will indeed be useful for us in the future, but we will not explore it in detail now.[20] We note that there are other perspectives to take simply because in mathematics there is always more than one perspective that one can take regarding certain ideas, and we should always be encouraged to find new perspectives as they always provide deeper insight into the ideas themselves.

In this chapter we have explored one such perspective on what we are "really" doing when we draw lines on a plane. Despite this being a subject of either middle school or high school mathematics, we should reflect on how far we have truly come. For prior to this, we had no perspective at all — we were just drawing lines on a

---

[19] Also for reasons that we will not discuss.

[20] Thus, if the previous couple of sentences have been mysterious, that is entirely okay because we will be going into much more detail in a later volume.

page with no idea of what we were "really" doing.[21] Now, we are able to precisely state what we are doing, and we can use the **much more general, abstract, and precise** language of sets and subsets. As we will see throughout these volumes, **all** of the math that we have come to know and love[22] throughout middle school and high school can, and often should, be rephrased in these more abstract terms. And again, it is important to keep in mind that this is **not** just more abstract machinery for the same old middle school math. Indeed, these abstractions have led us into the amazing world of logical paradoxes and infinitely high towers of infinities — it just so happens that middle school math is **also** included as a small subset of all of this as well.

We now hopefully have a better grasp of just how small of a portion of the mathematical world is taken up by this more "conventional" math. We have seen how a large part of this conventional math (namely, plotting lines on a plane) really only comes down to considering **particular** subsets of a **particular** set. However, the machinery that we have developed so far has applied to **all possible** sets — even those that have nothing to do with numbers. This is a truly significant change in perspective, as it is the first time that we are seeing a direct connection between this new world of abstract math and the world of math that we know from school. Accordingly, it is the first time that we are directly experiencing just how small the world of "school mathematics" is compared to the mathematical world that mathematicians **actually** deal with regularly. This is a paradigm shift in our understanding of the mathematical world and it deserves some reflection.

Finally, we note that none of this is to say that the world of "school mathematics" is any less important than any other field of math — we are here simply to expose ourselves to the rest of the mathematical world. Indeed, in the subsequent volumes we will introduce and explore mathematical structures that are incredibly diverse and profoundly beautiful. They will all have their roots in the work that we have done in this volume, yet they will end up taking a life of their own due to the tremendous flexibility of sets. Thus, the visi-

---

[21] At least, this is how the author felt when first learning about such things.
[22] Or possibly not love.

A FRESH PERSPECTIVE ON LINES

ble mathematical universe will continue to grow to expanses that we likely never thought was possible. This is all awaiting us in the future volumes of this series, but for now let us turn our attention to one final proof that we must give before ending this volume. It is a foundational and culturally necessary chain of reasoning for anyone who wants to learn about this new world of abstract math, and seeing as we have already made it this far we clearly meet this requirement.

Chapter 21

The Necessity Of Irrationality

A Bit More Notation

The current chapter concludes this first volume of our series. It has nothing to do with the construction of new sets from old ones, and in this sense may therefore seem somewhat out of place. However, it is only now that we have the mathematical sophistication to be able to fully appreciate that which we will learn in this chapter. Additionally, in the next volume we will start exploring very different ideas altogether, and so it is best for us to cover the ideas in this chapter now. Namely, in this chapter we will see that the real numbers (i.e., the elements in the set $\mathbb{R}$) that we have introduced and explored in several chapters throughout this volume are indeed, well, "real."

Before we discuss what we really mean for the real numbers to be "real" in the above sense, let us first introduce some more notation that will end up being useful for us in the future. In Chapter 17 we introduced the symbol "$\subseteq$" to denote the fact that one set is a subset of another, since we have agreed that the symbols "$B \subseteq A$" denote the statement that B is a subset of A. We noted that this symbol is analogous to the "$\leq$" symbol that we use to compare two **numbers** (as opposed to sets), so that if a and b are two numbers then $b \leq a$

A BIT MORE NOTATION

means that b is less than or equal to a. One of the most important analogies between the "$\subseteq$" and "$\leq$" symbols is the fact that, just as a number is less than or equal to **itself**, any set is a subset of **itself**. For example, we know that $5 \leq 5$ (because, in particular, 5 is equal to 5), and in a similar way we know that $A \subseteq A$ because — from the very definition of a subset — we know that any set is a subset of itself.[1]

Let us recall, however, that we can also compare numbers with the symbol "$<$." In particular, if we write "$a < b$," then we are saying that a is **strictly** less than b. By this we mean that a is **actually** less than b and **not** equal to it. Thus, for example, $4 \leq 5$ and $5 \leq 5$ are **both** true statements, as is the statement $4 < 5$. However, $5 < 5$ is **not** a true statement because 5 is **not** less than 5.

We now want to introduce the analogous symbol for sets — namely, a symbol to denote the fact that one set is a subset of another set **and** denotes the fact that we **know** the subset is not equal to the parent set. The obvious symbol to use for this is "$\subset$," so that we have the symbols "$\subseteq$" and "$\subset$" for comparing sets, in direct analogy with the symbols "$\leq$" and "$<$" for comparing numbers. Thus, when we write "$B \subset A$," we mean that B is a subset of A and that B is **not equal** to A. From the very definition of a subset, then, we see that this means that every element in B is also in A, but that there is at least one element in A that is **not** in B.

As an example, let us consider the sets $A = \{a, b, c, d\}$ and $B = \{a, b, c\}$. It is then the case that $B \subseteq A$ and $A \subseteq A$ are **both** true statements. Additionally, we see that $B \subset A$ because every element that is in B is **also** in A (so that B is a subset of A), but also that B is

---

[1] This is a good time to make the following important note. We can truthfully say that $4 \leq 5$ because indeed, 4 is less than or equal to 5. However, if we define the sets $A = \{5\}$ and $B = \{4\}$, we can **neither** say that $B \subseteq A$ **nor** that $\{4\} \leq \{5\}$. The reason for this is that when we write $4 \leq 5$, we are comparing 4 and 5 as **individual elements**. However, the satement "$B \subseteq A$" is a statement about **sets**, and it is indeed **not** true that B is a subset of A. Similarly, we cannot write "$\{4\} \leq \{5\}$" because again, the symbol "$\leq$" is not defined to compare sets — only numbers. It is very bad practice to forget what kinds of objects we are comparing, and it is extremely important that this notation is not abused. Thus, the statement "the only element in the set B is a number that is less than or equal to the only element in the set A" is true, however **none of** the expressions "$B \subseteq A$" **nor** "$\{4\} \leq \{5\}$" **nor** "$B \leq A$" denote this statement.

not equal to A because B does not have the element d. However, it is **not** true that $A \subset A$ because this would imply that A is not equal to A, which is obviously false. This is completely analogous to how we can not write $5 < 5$ even though we **can** write $5 \leq 5$. It is therefore always safer to use the "$\subseteq$" symbol, because if we know that one set is a subset of another set then this symbol is always applicable — in order to use the "$\subset$" symbol, however, we need the **extra** information that the subset is indeed genuinely missing some elements from the parent set. It is useful to give a name to the subsets of a set that are **not** equal to the larger set, so let us turn this into a formal definition.

Definition 21.1. Let A be a set. A subset B of A is called a **proper subset** of A if B is a subset of A and is not equal to A. If B is a proper subset of A, we denote this by $B \subset A$.

With this definition we can now rephrase some of the things that we have said above. Namely, we can say that the set $\{a, b, c\}$ is a proper subset of the set $\{a, b, c, d\}$, and we can say that no set is **ever** a proper subset of itself simply because any set always equals itself. Another interesting thing to note is that the empty set is a proper subset of any **non-empty** set. To see this, we recall that the empty set is a subset of **every** set. However, in order for the empty set to be a **proper** subset of some given set, it must not equal that given set. Thus, only if the given set is not the empty set itself is it the case that the empty set forms a **proper** subset.

Let us now turn our attention to seeing what all of this has to do with the set of real numbers.

A Chain Of Proper Subsets

Having introduced the notion of a proper subset in the previous section, we can now go on to relate the various sets of numbers that we have introduced so far in this volume. Namely, we recall that we have considered the set of natural numbers, which we denote by $\mathbb{N}$, and which consists of the elements $\{1, 2, 3, 4, ...\}$. We also have the set of all positive and negative whole numbers, including zero. This is nothing but the set that we considered in Chapter 10 as the set

A CHAIN OF PROPER SUBSETS

which has "infinity times 2" elements,[2] only now we are including the number zero. This set is often called the set of all **integers**, so that an integer is any whole number (positive, negative, or zero). The standard way to denote the set of all integers is with the symbol $\mathbb{Z}$. Thus we have that

$$\mathbb{Z} \equiv \{..., -3, -2, -1, 0, 1, 2, 3, ...\},$$

where the "..." on the left represents the fact that this pattern goes off to negative infinity, and the "..." on the right represents the fact that this pattern goes off to positive infinity.

Next, we recall that we have introduced the set of all fractions (which can also be positive, negative, or zero). Until now we have denoted this set by $FRAC$, but the more standard way of denoting this set is with the symbol $\mathbb{Q}$. We will therefore now start to conform to this convention. We then recall that the set $\mathbb{Q}$ consists of all possible expressions of the form $\frac{a}{b}$ where a and b are both integers (i.e., elements in $\mathbb{Z}$). We also must recall that we consider two of these expressions to be the same if one can be obtained from the other by multiplying the top integer and the bottom integer by the same number. For example, we have that

$$\frac{5}{10} = \frac{1}{2} = \frac{36}{72}$$

because $\frac{5}{10}$ can be obtained from $\frac{1}{2}$ by multiplying the top and the bottom of $\frac{1}{2}$ by 5, and $\frac{36}{72}$ can be obtained from $\frac{1}{2}$ by multiplying the top and the bottom by 36. We often call an element of $\mathbb{Q}$ a rational number, so that "rational number" is quite literally nothing but a fancy word for "fraction."

Lastly, we have the set of all real numbers, which we denote by $\mathbb{R}$. We can think of this set as "the continuum" of numbers, and we can represent it by drawing the number line as we did in the previous chapter. Equivalently, we can think of this set as the set of all numbers that can be expressed in the form

$$B.b_1 b_2 b_3 b_4 ...$$

---

[2] Recall, though, that we proved in that chapter that this set actually has the same cardinality as $\mathbb{N}$.

where B is an integer and where each b_i is some digit between 0 and 9. All of these b_i's simply form the "fractional part" of a given real number, and all we are saying here is that a real number is any number that can be expressed as a (possibly infinitely long) decimal. Thus, the number 5 is a real number with $B = 5$ and with all of the b_i's equal to zero, since 5 is already a whole number and therefore has no fractional part. Similarly, the number -4.6 is a real number with $B = -4$ (since this is the "integer part" of -4.6), with $b_1 = 6$ and with all of the **other** b_i's equal to zero. The number π is a famous real number because the decimal representation of the number π never ends and never repeats. Indeed, an approximate value of π is the famous 3.14, however an even better approximation for π is 3.14159, and an even better approximation[3] for π is 3.141592653. However, the full decimal representation of the number π never comes to an end, nor does it ever start to repeat itself. Thus, the number π is a real number with $B = 3$ as its integer part and has an infinitely long fractional part.[4]

Let us now begin to relate the sets $\mathbb{N}$, $\mathbb{Z}$, $\mathbb{Q}$, and $\mathbb{R}$ to each other, asking ourselves which sets are contained in other sets. Namely, let us begin with $\mathbb{N}$ and $\mathbb{Z}$. Since $\mathbb{Z}$ is the set of all whole numbers that are positive, negative, or zero, and since $\mathbb{N}$ is the set of only the **positive** whole numbers, we immediately see that $\mathbb{N}$ is a subset of $\mathbb{Z}$. Additionally, since $\mathbb{N}$ does not contain the negative whole numbers (or the number zero), we see that $\mathbb{N}$ is indeed a **proper** subset of $\mathbb{Z}$. We can therefore write that $\mathbb{N} \subset \mathbb{Z}$.

Let us now recall that any number that is in $\mathbb{Z}$ is also a fraction. This is because we can view, for example, the number 5 as a fraction which has the number 1 as its denominator.[5] Thus, the integers (i.e., elements of $\mathbb{Z}$) form a subset of the rational numbers (i.e., elements of $\mathbb{Q}$). However, there are lots of rational numbers, or fractions, which

---

[3]This is why, in America where the month is listed before the day when one writes the date, the time 9:26 and 53 seconds on March 14th, 2015, was such a big deal for lovers of π. Namely, at this time, it was 3/14/15 9:26:53.

[4]Indeed, using the above notation $B.b_1b_2b_3b_4$... for real numbers, we have for π that $B = 3$, $b_1 = 1$, $b_2 = 4$, $b_3 = 1$, $b_4 = 5$, $b_5 = 9$, and so on. We would be here for an infinite amount of time if we tried to write down every single b_i, however.

[5]This was discussed in more detail in Chapter 11 in case we need a quick review of these ideas.

A CHAIN OF PROPER SUBSETS

are **not** in $\mathbb{Z}$. For example, the number $\frac{1}{2}$ is a fraction and is therefore in $\mathbb{Q}$, but it is **not** in $\mathbb{Z}$ since $\mathbb{Z}$ contains only whole numbers (which are either positive, negative, or zero). Thus $\mathbb{Z}$ is a subset of $\mathbb{Q}$ which is **not** equal to $\mathbb{Q}$ and therefore $\mathbb{Z}$ is a proper subset of $\mathbb{Q}$. We can therefore write that $\mathbb{Z} \subset \mathbb{Q}$. Now, since we also know that $\mathbb{N} \subset \mathbb{Z}$, we can write that $\mathbb{N} \subset \mathbb{Z} \subset \mathbb{Q}$.

This is sometimes called a "chain of inclusions." For if a set C is contained in a set B, and if the set B is also contained in a set A, then we immediately know that C is contained in A. This is because if C is contained in B then every element in C is an element of B. And if B is contained in A then every element in B is also in A. Thus, since every element in C is in B and since every element in B is in A, it must be the case that every element in C is in A — and this is the definition of C being a subset of (or contained in) A.[6]

We then immediately see that if C is a **proper** subset of B, and if B is a **proper** subset of A, then C must also be a **proper** subset of A. For if C is a proper subset of B then every element of C is an element of B but there is at least one element of B that is not an element of C. And if B is a proper subset of A then every element of B is an element of A but there is at least one element of A that is not an element of B. Thus, since every element of B is an element of A and every element of C is an element of B, it must be the case that every element of C is an element of A. However, since there is at least one element of B that is not in C (since C is a proper subset of B) and since these elements must also be in A (since B is a subset of A), we have that there is at least one element of A that is not in C (namely, **at least** those elements in B that are not in C). Thus, C must be a proper subset of A.

The first half of the previous paragraph can be summarized as follows. If $C \subseteq B$ and if $B \subseteq A$, then $C \subseteq A$. This sentence reads word-for-word as "if C is a subset of B and if B is a subset of A, then C is a subset of A." We can then write the relationship between the sets A, B, and C as

$$C \subseteq B \subseteq A,$$

---

[6]This is indeed the exact argument given in the solution to Exercise 2 of Chapter 2.

and we can call this a "chain of inclusions" or a "chain of set inclusions." The second half of the previous paragraph can similarly be summarized as follows. If $C \subset B$ and if $B \subset A$, then $C \subset A$. This sentence reads word-for-word as "if C is a proper subset of B and if B is a proper subset of A, then C is a proper subset of A." We can then write the relationship between the sets A, B, and C as

$$C \subset B \subset A,$$

and we can also call this a "chain of inclusions" or a "chain of set inclusions."

We have already used this abstract machinery in the particular case that our sets are $\mathbb{N}$, $\mathbb{Z}$, and $\mathbb{Q}$, and since we had shown that $\mathbb{N} \subset \mathbb{Z}$ and that $\mathbb{Z} \subset \mathbb{Q}$, we can write $\mathbb{N} \subset \mathbb{Z} \subset \mathbb{Q}$. We now want to ask where the set $\mathbb{R}$ of real numbers sits in this chain of inclusions. In order to do this, we can immediately recall the fact that we stated (but did not, and will not prove) in Chapter 14 that **every fraction is a real number**. Namely, any number that can be expressed as a fraction can also be expressed as a (possibly infinitely long) decimal, and thus any fraction is a real number. This immediately tells us that $\mathbb{Q}$ is a subset of $\mathbb{R}$, and in symbols it tells us that $\mathbb{Q} \subseteq \mathbb{R}$. We now want to know if we can replace the "$\subseteq$" symbol with the "$\subset$" symbol. Namely, we want to know if there are elements in $\mathbb{R}$ that are **not** in $\mathbb{Q}$.

To motivate the fact that there **should** be elements in $\mathbb{R}$ that are not in $\mathbb{Q}$, let us recall that in Chapter 14 we showed that the cardinality of $\mathbb{R}$ is fundamentally larger than that of $\mathbb{Q}$. Thus, we know that there **must** be elements in $\mathbb{R}$ that are not in $\mathbb{Q}$ — otherwise, if every element of $\mathbb{R}$ were also an element of $\mathbb{Q}$, then $\mathbb{R}$ would be a subset of $\mathbb{Q}$, and since we already know that $\mathbb{Q}$ is a subset of $\mathbb{R}$ we would then know that $\mathbb{Q} = \mathbb{R}$, and it would then be the case that their cardinalities would have to be the **same**.[7] Since we have abstractly proved that the cardinalities of $\mathbb{Q}$ and $\mathbb{R}$ are **not** the same, it better be the case that the logic presented in the previous sentence is **not** applicable, and therefore it must be the case that there are elements in $\mathbb{R}$ that are not in $\mathbb{Q}$. In other words, it must be the case (in order

---

[7] For if two sets are **equal**, they certainly have the same cardinality.

A CHAIN OF PROPER SUBSETS   221

for all that we have discussed so far to remain true) that there exists a real number which is **not** a fraction.

It is important to note that the logic that we have presented so far is indeed a fully rigorous proof of the fact that $\mathbb{Q}$ is a **proper** subset of $\mathbb{R}$. Namely, we have **already proven** that the cardinality of $\mathbb{R}$ is different from (and larger than) the cardinality of $\mathbb{Q}$, and we **already know**[8] that $\mathbb{Q}$ is a subset of $\mathbb{R}$ — i.e., that every fraction is also a real number. In the previous paragraph we took these two facts alone and showed that there must be some elements in $\mathbb{R}$ that are not in $\mathbb{Q}$. However, this is all very abstract. We have been able to prove the existence of some object (namely, a number that is in $\mathbb{R}$ but not in $\mathbb{Q}$) without **explicitly** constructing or describing that object. This is a very interesting and significant phenomenon that often arises in mathematics. Namely, if we simply want to show that such a number exists, then we would be done. However, if we wanted to write such a number down, or get to know the details of such a number, then we are not done — our abstract proof of the **existence** of such a number does nothing for us if we actually want to examine a particular number. Thus, in the remainder of this chapter, we will not concern ourselves with **whether or not** $\mathbb{Q} \subset \mathbb{R}$ — i.e., with whether or not there exists at least one real number that is **not** a fraction — simply because we already know that this is true, but rather we will focus on the task of **finding** such a number and **proving** that it **is** such a number.

In particular, we will show that the square root of two,[9] which we will denote by the symbol $\sqrt{2}$, cannot be written as a fraction. Before showing this, however, it is enlightening to reflect upon whether or not the number $\sqrt{2}$ is "really" a number. Namely, the definition of

---

[8] We have not proven this fact, but we can rest assured that it has indeed been proven by others.

[9] I.e., that number which, when multiplied to itself, gives the number 2. For example, the square root of 16 is 4, since 4 multiplied by 4 is 16. Similarly, the square root of 25 is 5, since 5 times 5 is 25. However, there is no **whole** number that, when multiplied to itself, gives 2. This is because 1 times 1 is 1, and 2 times 2 is 4. Thus, the number which multiplies to itself to give 2 must lie somewhere between 1 and 2. In particular, a good approximation to $\sqrt{2}$ is 1.41421356, though the exact value of this number is given by a decimal that is infinitely long and that never sets up any pattern.

real numbers as possibly infinitely long decimals may seem somewhat artificial, whereas fractions seem much more "natural." Thus, if we claim that there is a number that is in $\mathbb{R}$ but is **not** a fraction, we might get the feeling that this number is rather artificial as well. Therefore, let us now show that the square root of two is indeed absolutely necessary in the mathematical world, and not at all a human construct.

One way to see that $\sqrt{2}$ is indeed a completely necessary mathematical object is to consider a right triangle like that shown in Figure 21.1. Namely, if we let the side lengths of a and b both[10] equal 1,

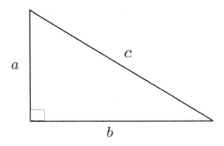

Figure 21.1: Right triangle with sides labeled a, b, and c.

and if we apply the famous and likely familiar theorem of Pythagoras which says that
$$a^2 + b^2 = c^2,$$
then we find that $1^2 + 1^2 = c^2$. This comes from simply plugging in our specified values of $a = 1$ and $b = 1$ into the above Pythagorean theorem. Now, 1^2 is simply notation for "1 times 1," which is 1. Thus, plugging in 1 for 1^2, we have that $c^2 = 1 + 1$ and therefore we have that $c^2 = 2$. In other words, we have that c is a number which, when multiplied to itself, gives the number 2. Therefore, we have just shown (using nothing by Pythagoras' amazing (and true) theorem) that a right triangle with two sides of length 1 will have a third side of length $\sqrt{2}$. We certainly want to have a system of numbers that can describe a geometric object as basic as a right triangle with two

---

[10]Note that, in this case, Figure 21.1 is not drawn to scale, since in that figure the length of a is not the same as that of b.

THE PROOF

sides of length 1, and therefore the number $\sqrt{2}$ has a very real and necessary existence.

Having established that the number $\sqrt{2}$ is a mathematical entity that cannot be ignored, we will now show that $\sqrt{2}$ cannot be written as a fraction, and is therefore irrational.[11] The fact that $\sqrt{2}$ is irrational — a fact that we will show in the next section — gives credence to the idea that real numbers, despite their strange and unintuitive expressions as possibly infinitely long decimals, are indeed very "real."

The Proof

We now set out to prove that $\sqrt{2}$ is not a fraction. As in most proofs of "negative" statements,[12] our most powerful weapon is that of proof by contradiction.[13] Thus, we suppose that $\sqrt{2}$ **is** a fraction and we will then derive "true" statements that follow from this supposition until we arrive at a conclusion that contradicts previously established truths. This will then show us that our supposition was wrong — namely, that $\sqrt{2}$ is **not** a fraction.

Once we suppose that $\sqrt{2}$ is a fraction, then we can write $\sqrt{2} = \frac{a}{b}$ for some a and b where both a and b are integers. This is simply because **any** fraction can be written as $\frac{a}{b}$ for some a and b where both a and b are integers, and therefore once we suppose that $\sqrt{2}$ is a fraction then we can write it this way as well. It is important to note that we have no idea what the precise form of this fraction is, and it is even more important to note that it does not matter what

---

[11] Namely, an irrational number is — by definition — any number which is not rational. Since rational numbers are nothing but fractions, an irrational number is nothing but a number which is **not** a fraction. All of this is just fancy terminology that mathematicians have introduced for various reasons, and the important thing to take away here is that rational numbers and fractions are completely identical (different words for the same idea), and that irrational numbers are all the numbers which are not rational numbers.

[12] By a "negative" statement we mean a statement that some mathematical structure does **not** have some property. Namely, the statement we are concerned with here is a "negative" statement because we want to show that $\sqrt{2}$ does **not** have the property of being a fraction.

[13] This method of proof was discussed in more detail in Chapter 13.

the precise form of this fraction is. The only important thing here is that **if** $\sqrt{2}$ is a fraction, then we can write $\sqrt{2} = \frac{a}{b}$ for **some** a and b. We can further suppose that $\frac{a}{b}$ is already "fully reduced," meaning that there are no common factors between a and b. For example, the fraction $\frac{1}{2}$ is fully reduced because 1 and 2 have no common factors between them, however the fraction $\frac{6}{8}$ is **not** fully reduced because we can divide both 6 and 8 by 2. In this case, the "fully reduced" form of $\frac{6}{8}$ is $\frac{3}{4}$, which is obtained by dividing the top and bottom by the common factor of 2. We see that **any** fraction has a fully reduced form that it is equivalent to, and all we are supposing here is that the fraction $\frac{a}{b}$, which we are supposing is equal to $\sqrt{2}$, is already fully reduced. There is nothing stopping us from doing so.

There are now four facts that we need in order to find a contradiction from our supposition that $\sqrt{2} = \frac{a}{b}$.

Fact 1. An odd number multiplied by an odd number is again an odd number, and an odd number multiplied by an even number is an even number. To see this, we note that an odd number **plus** an odd number is an even number. This is because an odd number is one away from an even number, so that when we add two odd numbers to each other we are adding two numbers which are **both** one away from an even number. The result is therefore a number which is two away from an even number, but any number which is two away from an even number is also an even number. For similar reasons, an odd number plus an even number is again an odd number.

It is useful to see a particular instance of all of this. For example, 7 and 9 are both odd numbers, and 7 plus 9 is 16, which is an even number. To see why this is a general phenomenon, we can think of 7 as being 6 plus 1 (namely, being one away from an even number), and we can think of 9 as being 8 plus 1. Then 7 plus 9 is actually 6 plus 8 plus 1 plus 1. The sum of two even numbers is again even, so that 6 plus 8 is even and the sum of 1 and 1 is 2, so that our final result is also even.

We therefore see that an odd number can be thought of as an even number with a "remainder" of 1. Thus, if we add two odd numbers together, then we are adding two even numbers as well as two "remainders" of 1, leaving us with an even number. Similarly, if

THE PROOF

we add an odd number to an even number, we are still left with only one "remainder" of 1, and therefore with an odd number.

Let us now reconsider **multiplying** an odd number by an odd number. Namely, we note that multiplying an odd number by an odd number is nothing but adding an odd number to itself an odd number of times. For example, 7 times 5 is exactly the same as adding 7 to itself 5 times. Thus, we are adding an even number with a "remainder" of 1 to itself an odd number of times, so that we have an odd number of "remainders" of 1, which leaves us with an odd number. This indeed works out when we look at specific examples, since 7 times 5 is 35 which is again an odd number.

In a similar way, an odd number multiplied by an even number is again an even number, but this is even easier to see. The reason for this is that an **even** number added to itself any number of times is even. For example, adding 6 to itself any number of times will always give us back an even number. Thus, since we can think of an odd number multiplied by an even number as an even number added to itself an odd number of times, we immediately have that an odd number times an even number is an even number.

To recap, the first fact that we need is that **an odd number times an odd number is an odd number, and an odd number times and even number is an even number.**[14]

Fact 2. An even number a can always be written as 2 times some other number, where this other number can be either even or odd. This is, in fact, just the definition of a number being even — namely, that it can be written as 2 times some other number. Thus, if a is an even number, then $a = 2k$ for some integer k. As an example, we have that 6 is even and indeed 6 equals 2 times 3. Thus, in this case, if $a = 6$ we have that $k = 3$. Similarly, 28 is an even number and can be written as 2 times 14. Thus, in this case, $a = 28$ and $k = 14$. To recap, **any even number a can be written as $a = 2k$ for some other number k.**

Fact 3. If $\frac{a}{b}$ is a fraction that is "fully reduced" in the sense that

---

[14]Indeed, an even number times an even number is again an even number, but this fact will not be useful to us in our proof.

we described above, then it cannot be the case that both a and b are even. This is because if a and b are both even, then they both have a common factor of 2 (using Fact 2), and this contradicts the fact that $\frac{a}{b}$ is fully reduced. Thus, since **every** fraction can be fully reduced, we know that we can reduce any fraction to the form $\frac{a}{b}$ where at **most** only one of a or b is even.[15] To recap, we have that **once a fraction is fully reduced, it cannot be the case that both the numerator and denominator are even**.

Fact 4. Any integer is either even or odd — never both and never neither. The integer 8 is even and not odd, the integer -5 is odd and not even, and the integer 0 is even and not odd. To recap, **no integer can be neither even nor odd, and no integer can both even and odd** — it is **always** one or the other.

With these four facts at our disposal, we can now return to our proof. Let us recall that we have supposed that $\sqrt{2} = \frac{a}{b}$, and once we make this supposition then we know that we can also suppose that $\frac{a}{b}$ is fully reduced, since any fraction at all can be fully reduced. We will soon find that the supposition that $\sqrt{2} = \frac{a}{b}$ leads to a contradiction to the above facts, thus disallowing us from making such a supposition and therefore proving for us that indeed $\sqrt{2}$ is not a fraction. Let us see how this contradiction arises.

We note that if $\sqrt{2} = \frac{a}{b}$, then it must be the case that multiplying $\sqrt{2}$ by $\sqrt{2}$ is the same as multiplying $\frac{a}{b}$ by $\frac{a}{b}$. This is simply because we have supposed that $\sqrt{2}$ and $\frac{a}{b}$ are the same number, and therefore multiplying them to themselves should give the same number as well. Now, $\sqrt{2}$ times $\sqrt{2}$ is simply 2, since $\sqrt{2}$ is by definition the number that, when multiplied to itself, gives the number 2. Additionally, we have that $\frac{a}{b}$ times $\frac{a}{b}$ is nothing[16] but $\frac{a^2}{b^2}$. We have therefore shown, after equating the product of $\sqrt{2}$ with itself and the product of $\frac{a}{b}$

---

[15]There are many fractions, like $\frac{5}{7}$, which are fully reduced and where **neither** the top nor the bottom integers are even. What we have noted in this fact is that once a fraction is fully reduced, then it must be the case that **either** the numerator or the denominator (and possibly both) is odd.

[16]Recall that multiplying two fractions together is done by simply "multiplying the top by the top" and "multiplying the bottom by the bottom."

THE PROOF

with itself, that
$$2 = \frac{a^2}{b^2}.$$

Now, by simply "bringing b^2 to the other side," or equivalently by multiplying both sides by b^2, we find that

$$a^2 = 2b^2.$$

This means, by definition of an even number, that a^2 is even. This is simply because we have now found that a^2 is 2 times some other number, where in this case this "other number" is b^2. However, if a^2 is even, then it must be the case that a is even as well. For if a were **not** even, then by Fact 4 above it would have to be odd. And if a is odd then by Fact 1 above it would be the case that a^2 is odd as well, since a^2 would be an odd number multiplied by an odd number (if a is odd). Therefore, since a^2 is indeed even, it must be the case that a is also even.

Let us note that **solely** from the assumption that $\sqrt{2} = \frac{a}{b}$, we have been able to prove that a is even. Now, since a is even and using Fact 2 above, we know that a must be 2 times some other number. Let us call this other number k. Then we can write $a = 2k$. If we now plug $a = 2k$ into the equation

$$a^2 = 2b^2$$

that we found above, we find that

$$(2k)^2 = 2b^2$$

and therefore we have that

$$4k^2 = 2b^2.$$

This all follows by first replacing a with $2k$ and then noting that, in order to square a, we must square $2k$. Then, to square $2k$, we must multiply $2k$ to itself which means we must multiply 2 to itself **as well as** multiply k to itself. Thus, $a^2 = (2k)^2 = 4k^2$ where in the last equality the "squared" superscript only applies to the k since we have already squared the 2 to get 4.

We have now found that once we write $a = 2k$ (which we can do because we found above that a is an even number), then we have that

$$4k^2 = 2b^2.$$

Let us now divide both sides of this equation by 2. We can do this because if one equation is true, then it must still be true after we do the same operation to both sides. We therefore find that

$$2k^2 = b^2.$$

This means that b^2 is an even number because we see that it is 2 times some other number, where the "other number" is now k^2. Arguing in exactly the same way as we did for a, we see that if b^2 is an even number then b itself must be an even number, for if b is odd then it would be the case that b^2 is odd. Therefore, we have now also found that b is an even number.

Thus, from our initial supposition that $\sqrt{2} = \frac{a}{b}$, and performing nothing but **allowable** manipulations from this supposition, we have found that **both** a and b are even numbers. This contradicts Fact 3, however, because we initially assumed that $\frac{a}{b}$ was fully reduced and Fact 3 tells us that if $\frac{a}{b}$ is fully reduced then it **cannot** be the case that both a and b are even.

We have therefore found our contradiction, which means that our initial supposition was wrong, which means that $\sqrt{2}$ **cannot** be written as a fraction. This now concretely shows us that irrational numbers[17] are numbers that we have to confront if we wish to even ask about the third side of the triangle in Figure 21.1.

In fact, it turns out that there are **many** more irrational numbers than there are rational ones. This is perhaps not surprising, since this must be so in order for it to be possible that the cardinality of $\mathbb{R}$ is a fundamentally larger infinity than that of $\mathbb{Q}$, as we saw in Chapter 14. While it is not the case that **all** irrational numbers are as fundamental and necessary[18] as $\sqrt{2}$, indeed many of them are. For example, as we have mentioned before,[19] the number $\pi = 3.1415926...$

---

[17] Numbers which are **not** rational, i.e., numbers which are not fractions.

[18] In the sense of having a direct and simple geometric representation like that of a right triangle with two sides of length 1.

[19] Though we have not and will not prove this fact.

is indeed irrational, and it has the geometric meaning of being the area of a circle whose radius has length 1, or equivalently the number of diameters that can fit around the circumference of any circle.

We therefore must view irrational numbers as "existing" in the exact same way that the more intuitive notions of fractions and whole numbers exist. The sense in which negative numbers truly "exist" might be another cause for concern, and in the next volume we will build up a framework in which their existence becomes more concrete.

Outro and Intro

In the next volume we will go on to take the theory of sets that we have developed in this volume and use it to create a formalism in which many of our "familiar" mathematical structures exist in a much more abstract and precise way. We will introduce and study the mathematical structures known as groups, rings, and monoids, just to name a few. One consequence of this will be that we will be able to give a much more abstract and precise meaning to the set of natural numbers $\mathbb{N}$, the set of integers $\mathbb{Z}$, and the set of fractions $\mathbb{Q}$. We will also be able to describe the relationship between these sets in a more precise and useful way than simply discussing which sets are (proper) subsets of other sets.

This discussion of $\mathbb{N}$, $\mathbb{Z}$, and $\mathbb{Q}$ in the next volume is not, however, going to be our main purpose. Just as plotting lines on a plane is not the main purpose of rephrasing mathematics in the language of sets, the main purpose of introducing the structures of groups, rings, and monoids is not to be able to discuss integers and fractions. Indeed, the main purpose of rephrasing mathematics in the language of sets is to be able to discuss and describe more abstract and more general phenomena that occur in the realm of pure logic. As we will see, the main purpose of rephrasing algebra in terms of more abstract objects like groups and rings is to uncover a unity and beauty that underlies more familiar structures like numbers. This will give us a more powerful collection of mathematical tools that we can use to construct mathematical structures of increasing complexity and beauty.

It is in this process that the true beauty of math lies. Namely, by taking familiar mathematical structures and generalizing them to

OUTRO AND INTRO 231

more abstract structures, we uncover a new world of unfamiliar yet precise logical constructions that we can then begin to explore. Upon exploring this new world, we find remarkable and unexpected unifying patterns and connections between seemingly disparate logical structures, giving us a new and deeper understanding of the structures with which we began, and of the structures that we uncover along the way.

In this volume we began at the beginning, and found that the most general type of structures that we can describe are elements and sets. As we started exploring sets we found that there are indeed wildly different kinds of sets — sets of apples, sets of NBA players, infinite sets, finite sets, sets of numbers, and various combinations of all of these. We saw how we could take two sets and define functions from one to the other, and we saw how we could also take two sets and form a third set from them. We examined subsets of sets, sets of sets, and sets of subsets of sets. Perhaps most importantly we saw how the more familiar mathematical structures like numbers and lines on a plane fit in as a very tiny part of all of this abstract machinery, and that the abstract machinery gives us **much** more to explore than the familiar structures themselves would lead us to believe.

In the next volume we will uncover new types of unity that exist amongst large classes of sets, and we will do so by specifying new types of structure within sets. We will be able to give more precise and abstract meaning to the processes of addition, subtraction, multiplication, and division, and in doing so we will uncover brand new forms of these processes that exist within sets that (seemingly) in no way resemble sets of numbers. Sets will no longer be static, unchanging mathematical structures, but rather will be given life through various interactions of their elements amongst themselves. This different perspective on sets will give the familiar world of algebra a whole new meaning — one that is much more flexible, general, and diverse than one could ever have possibly imagined from middle or high school homework assignments which ask one to "solve for x." Therefore, let us now go on to see this whole new world of abstract algebra and turn our attention towards stepping up our set theory game by giving life to sets.

Solutions To Exercises

Solutions To Chapter 1

1) **Define a set, any set at all, in two different ways. For example,**
$$\{\text{all humans over the age of 35}\}$$
is the same set as
$$\{\text{all humans who are not 35 years of age or younger}\},$$
yet these two sets are defined in different ways. Similarly,
$$\{\text{those who think that LeBron is better than Kobe}\}$$
is the same set as
$$\{\text{those who enjoy being wrong}\}.$$

Solution: There is no single correct answer to this problem, and we have given a couple examples of answers in the statement of the problem itself. Since both of the examples in the problem statement involve sets whose elements are material things (namely, people), we will provide a couple more examples here involving sets of numbers. As one example, we can define the set $\{1, 2, 3, 4, 5\}$ in many different ways. First, we can simply write out its elements as we just did — namely, the set is defined to be $\{1, 2, 3, 4, 5\}$. As a second way of defining the same set, we could say that we want to consider the set of all whole numbers that are larger than zero and less than 6, as this also specifies the set $\{1, 2, 3, 4, 5\}$. As a third way of defining the same set, we could say that we want to consider the set of

SOLUTIONS TO CHAPTER 1

all whole numbers that, when multiplied by two, gives an element of the set $\{2, 4, 6, 8, 10\}$. It is then the case that all possible whole numbers that satisfy this condition are precisely the elements in the set $\{1, 2, 3, 4, 5\}$.

This gives three different ways of defining the exact same set. The problem only asks for two different ways, and indeed there infinitely many different ways of defining the same set. For example, we could define the set $\{1, 2, 3, 4, 5\}$ by saying that it is the set of all whole numbers that, when multiplied by three, gives an element in the set $\{3, 6, 9, 12, 15\}$. This gives a fourth way of defining the same set. Equivalently, we could also say that the set $\{1, 2, 3, 4, 5\}$ is the set of all whole numbers that, when multiplied by four, gives an element in the set $\{4, 8, 12, 16, 20\}$. By continuing this with multiplication by five, six, and so on, we see that there are infinitely many ways of defining the same set.

As a second example, let us consider the set $\{2, 4, 6\}$. We can define this set in many different ways, and any two of them would be sufficient for this problem. One way would be to simply write down the elements as we just did. Namely, writing $\{2, 4, 6\}$ completely defines our set. We can also define this set by saying that it is the set of all even numbers that are larger than 0 and less than 8, or equivalently by saying that it is the set of all even numbers that are larger than 0 and less than 7. Namely, there are only three even numbers that are both larger than 0 and less than 8, and they are the same as those even numbers which are larger than 0 and less than 7. We can also say that the set $\{2, 4, 6\}$ is the set of all numbers that can be obtained by multiplying the elements in the set $\{1, 2, 3\}$ by 2, or we can say that it is the set of all numbers that can be obtained by dividing the elements in the set $\{10, 20, 30\}$ by 5. Again, we see that there are infinitely many different ways to define the set $\{2, 4, 6\}$, and any two of them would be sufficient for this problem.

It is very important to note that any given set may have (and usually does have) many different ways in which it can be defined, because oftentimes in math certain sets are presented in various ways and we need to be able to extract the relevant information. Namely, if we were presented with the set of all numbers that can be obtained from the elements in the set $\{10, 20, 30\}$ by dividing by 5, it will be useful for us to identify this set simply as $\{2, 4, 6\}$ even though this simpler form is not how the set was initially given to us.

2) Try to define some crazy sets, and try to determine if two sets are the same or not. For example, can we define the set of all ideas? What about the set of all thoughts? What makes two different thoughts distinct? (Remember, elements in a set need to be distinct!) Is the set of all ideas equal to the set of all thoughts? It depends on how we define them! Let us not get too worked up about this stuff yet though — we are just trying to see that "mathematical thought" is indeed much broader than we could have ever imagined!

Solution: This exercise also does not really have a single solution since each one of us may think of wildly different examples. However, let us at least get the ball rolling here. Some crazy sets that one could define are the following: the set of all toenails (crazy just because it is gross), the set of all universes (each universe is a single element — is there more than one element? Are there infinitely many elements?), and the set of all sets! The set of all sets is simply the set whose elements are themselves sets, and moreover it contains **all** of the sets. We will discuss this wacky set a lot more in the solution to the next exercise as well as in Chapter 4.

The part of this exercise that asks about the difference between the set of all ideas and the set of all thoughts is primarily here to show us the importance of having rigor in our definitions. We need to **precisely** define what we mean by "a thought" and "an idea" before we can ask whether or not these two sets are the same. Let us make the following definitions (although it should be clear that there is no right or wrong definition, nor any exact, mathematical definition). A **thought** will be defined to be any reaction, feeling, emotion, opinion, observation, or any other mental process, either verbalized or not, by any sentient being anywhere. This is of course a complex issue because now we would have to define every term in that definition — namely, we would have to precisely define "feeling," as well as

SOLUTIONS TO CHAPTER 1   235

"emotion," as well as "sentience," etc. — but let us just go with this. An **idea**, on the other hand, will be defined to be any mental facility brought forth by any sentient being from the motivation to accomplish any kind of goal.

With these (highly inexact) definitions, it is clear that the set of all ideas and the set of all thoughts are not the same, because there are several thoughts that are not ideas. This is simply because — as we have defined things — there can be thoughts which are **not** created with the motivation to accomplish any goal, and such thoughts would not be ideas because ideas are required to be motivated by a desire to accomplish some kind of goal. It is the case, however, that any idea is indeed a thought, simply because an idea is a mental facility (of a particular kind), and a thought is any mental facility. For example, "I am hungry" is a thought that has been had. This thought is not an idea, however, because there is absolutely no goal that motivated its existence. We can be quite sure that "I should eat" is a thought that has also been had. This thought is indeed an idea, because it was motivated by wanting to accomplish the goal of no longer being hungry.

It cannot be over-emphasized that all of these statements are still highly imprecise and open for interpretation and/or debate. The point of this exercise is to see the importance of being able to make **rigorous and precise** definitions, and a consequence of this exercise is that we also get to see how difficult making such definitions is in regards to almost all non-mathematical structures. Namely, it is **much** easier to make rigorous, precise, and objective definitions about mathematical structures like numbers and geometric objects than it is to do so about non-mathematical structures like thoughts (or anything that is social, cultural, or subjective in any way).

3) Is there a set that contains itself as an element? (This one is tricky and it is something we will be addressing more in Chapter 4, so we shouldn't panic if this exercise eludes us right now.)

Solution: The answer is yes, there is such a set, and one example of a set that contains itself as an element is actually in the solution to the previous exercise. Namely, after some thought, we notice that **the set of all sets** contains itself as an element! It is, after all, a set, and since it is the set of all sets, it must have **itself** as an element.

As another example[1] of a set that contains itself as an element, we can consider "the set of all sets that do not contain the element 2." This is the set that has other sets as its elements. Namely, for any set in the world, we ask whether or not it contains the element 2. If the given set does **not** contain the element 2, then that set **is** an element of "the set of all sets that do not contain the element 2," and if the given set **does** contain the element 2, then this set is **not** an element of "the set of all sets that do not contain the element 2." In order to see if "the set of all sets that do not contain the element 2" contains **itself** as an element, we must ask if this set contains the element 2. Indeed, this set does **not** contain the element 2, because this set only contains "sets that do not contain the element 2." Namely, the element 2 is a number, and so it is **not** a "set that does not contain the element 2." Thus, the element 2 is **not** in the set "the set of all sets that do not contain the element 2," and therefore "the set of all sets that do not contain the element 2" contains **itself** as an element, simply because it is **itself** a set that does not contain the element 2. Using the exact same logic, we see that "the set of all sets that do not contain x" is a set that contains itself as an element, regardless of which object x we choose.

On the face of it there is nothing logically wrong with a set containing itself as an element, but as we explore sets of sets more deeply in Chapters 3 and 4 we will find that there is indeed something that goes **very** wrong when there exists sets that contain themselves as elements. Nonetheless, "the set of all sets" as well as the kinds of sets that we discussed in the previous paragraph work to solve this

---

[1] As a warning to the reader, the current paragraph may appear extremely confusing upon first (and possibly even second and third) reading. That is entirely okay. The reader is encouraged to give this paragraph multiple readings, and to read each phrase carefully and slowly. However, the reader should also rest assured that we will be taking a closer look at these issues in the next couple of chapters and thus, if this paragraph does not make perfect sense right now, it hopefully will soon. So let us read on and possibly return to this exercise once we are a bit more familiar with sets, and sets of sets.

SOLUTIONS TO CHAPTER 2

problem and so we will allow them, for now. If some of this does not make all the sense in the world at the moment, let us not worry, because we will be addressing these issues more slowly in the coming chapters.

Solutions To Chapter 2

1) (a) Write down four subsets of the set

$$\{a, b, c, d, e, f, g\}.$$

(b) Write down three subsets of the set

$$\{\text{Kobe}, \text{LeBron}, 8, \text{love}, \text{fruit loops}, 27\}.$$

Solution: (a) We will write down more than four subsets of $\{a, b, c, d, e, f, g\}$, though we will not write down all of the **possible** subsets. One subset that we always know we have is the empty set, since the empty set is a subset of **every set**. We also know that $\{a, b, c, d, e, f, g\}$ is a subset of itself, since **any set** is a subset of itself. That is already two subsets of $\{a, b, c, d, e, f, g\}$. We then have 1-element subsets like $\{a\}$, $\{b\}$, $\{d\}$, or $\{g\}$, as well as 2-element subsets like $\{a, b\}$, $\{a, c\}$, or $\{d, g\}$. We also have 3-, 4-, 5-, and 6-element subsets, though we will not list an example of each. As an example of a 5-element subset of $\{a, b, c, d, e, f, g\}$, we could use $\{b, d, e, f, g\}$. Any four of the subsets that we have written so far would do the job that was asked of us, and there are many, many other possibilities as well. Indeed, since there are 7 elements in $\{a, b, c, d, e, f, g\}$, we can use what we learned in this chapter to see that there are $2^7 = 128$ possible subsets of $\{a, b, c, d, e, f, g\}$, so we do not recommend writing all of them down explicitly!

(b) The set

$$\{\text{Kobe}, \text{LeBron}, 8, \text{love}, \text{fruit loops}, 27\}$$

is a 6-element set and therefore there will be 64 possible subsets. We will therefore only list a few, and any three of them will be sufficient.

As always, we know that the empty set is a subset of

$$\{\text{Kobe}, \text{LeBron}, 8, \text{love}, \text{fruit loops}, 27\},$$

and we also know that

$$\{\text{Kobe}, \text{LeBron}, 8, \text{love}, \text{fruit loops}, 27\}$$

is a subset of

$$\{\text{Kobe}, \text{LeBron}, 8, \text{love}, \text{fruit loops}, 27\}.$$

We can then choose any 1-element subset, like {Kobe}, or {8}, or {love}, or we could choose any 2-element subset, like {love, fruit loops} or {Kobe, 8}. We could also just as well choose any other 2-, 3-, 4-, or 5-element subset, though we will end our discussion of subsets of this set here.

2) If C is a subset of B, and B is a subset of A, then C is a subset of A. True or false?

Solution: True. Recall that in order to see if C is a subset of A, we need to ask if every element in C is also in A. Now, we know (because it is given in the problem statement) that B is a subset of A, so we therefore know that every element in B is also in A. We also know that C is a subset of B, so we know that every element in C is an element in B. Thus, if we are handed an element of C, then we know it is in B, and since it is in B, we then know it is in A. Since this is true for **any** element that we were handed from C, we have thus established that **every** element in C is also in A. In other words, we have established that C is a subset of A.

This result may seem obvious (or it may not), but it is a good example of what it means to be rigorous and detailed. Namely, it is not explicitly written in the definition of a subset that if C is a subset of B and if B is a subset of A then C is a subset of A. We therefore need to explicitly prove this statement no matter how obvious it may or may not seem, and this is what we have done in this exercise.

SOLUTIONS TO CHAPTER 2

3) It is hopefully clear that there are infinitely many 1-element sets "out there." This is because we can form the sets $\{1\}, \{2\}, \{3\}, \{4\}, \{5\}$, and so on, so this already gives us infinitely many 1-element sets (and this is not even including $\{\text{cup}\}, \{\text{table}\}, \{\text{this chair}\}, \{\text{that chair}\}, \{\text{that chair over there}\}, \{\text{that other chair}\}, \{\text{lion}\}$, and so on). Similarly, there are infinitely many 2-element sets, and infinitely many 3-element sets (and so on). The question is, then, how many 0-element sets are there? (Hint: We gave away the answer somewhere in this chapter, so now we need to understand why this is so.)

Solution: There is exactly one 0-element set, and it is precisely the empty set that we have already introduced. In other words, all 0-element sets are the same, and they are all equal to the empty set. To show that this is the case, we employ a very common method of proof. Namely, we will show that there is (at least) one 0-element set, and then we will suppose that there is another. We will then show that this supposed "other" is in fact the **same** as the original one, and that therefore there can only be one 0-element set.

We know that the empty set is a 0-element set, simply because that is how the empty set is defined. Thus we know that there is **at least** one 0-element set. Now let us suppose that there is another 0-element set. We want to show that any other 0-element set **is actually equal to** the empty set, thus showing that the empty set is the **only** 0-element set. How do we show that two sets are equal? We show that they have the exact same elements. Namely, we show that every element in one set is in the other, and every element in the other is also in the first. I.e., we show that they both "contain each other," thus showing that they are equal. In this proof we will rely heavily on the validity of **vacuous truths**, which we examine more thoroughly in the first appendix of this volume.[2]

Let us denote the empty set by $\emptyset$ and let us call the other 0-element set that we are supposing exists A. We want to show that $\emptyset = A$, for this would then mean that **any** 0-element set is equal to the empty set and therefore that the **only** 0-element set is the

---
[2] Now would be a good time to familiarize ourselves with this concept by reading Appendix A if we are unfamiliar with vacuous truths.

empty set. To do this — namely, to show that $\emptyset = A$ — we need to show that every element in $\emptyset$ is also in A, and vice versa. Now, there are no elements in $\emptyset$, so every element in $\emptyset$ is vacuously in A. In the exact same way — namely using the vacuously true statement that a 0-element set is a subset of every set — we conclude that every element in A is also in $\emptyset$. Thus these two sets are equal, and so there is only one 0-element set!

This is a much more detailed discussion than is perhaps necessary for this problem, but it is a nice setting for discussing this type of logic, which will be relied upon heavily in future chapters and volumes to derive much less trivial results.

Solutions To Chapter 3

1) Let $A = \{1, 2, 3, 4, 5\}$ and let $B = \{\{1,2\}, \{3,4\}, 5\}$. How many elements does A have? How many elements does B have? Does $A = B$?

Solution: The set A has 5 elements, and the set B only has 3! The result for A should be clear, since its elements are 1, 2, 3, 4, and 5. The result for B follows because the elements in B are the **sets** $\{1, 2\}$ and $\{3, 4\}$, as well as the number 5. Namely, $\{1, 2\}$ and $\{3, 4\}$ are **single elements** in the set B — they are each sets in their own right, but they are **single elements** of B. Accordingly, A and B are **not** equal as sets. We can see this many ways. First of all, a set with 3 elements can never be equal to a set with 5 elements. More concretely, we see that the element 1 is not in B, nor are the elements 2, 3, or 4, yet all four of these elements are in A. We note that 1 **is** an element of $\{1, 2\}$ and that $\{1, 2\}$ is an element of B, but this does **not** mean that 1 is an element of B. Similarly, the element $\{1, 2\}$ is not in A, nor is the element $\{3, 4\}$, yet both $\{1, 2\}$ and $\{3, 4\}$ are elements of B. When viewed in this way, it becomes clear that these sets are not equal.

2) Write down (in the bracket notation that we have developed) all of the elements in the set $P(A)$ when $A = \{$Kobe, LeBron, MJ$\}$.

SOLUTIONS TO CHAPTER 3 241

Solution: We recall that for any set A, $P(A)$ is the set of **all subsets** of A. Namely, $P(A)$ has for its **elements** all of the possible subsets of A. Let us now see how this works for the specific set $A = \{\text{Kobe, LeBron, MJ}\}$. We know that the empty set is a subset of every set, so that if we denote the empty set by $\emptyset$ then we know that $\emptyset$ is an **element** of $P(A)$. Let us now move on to the 1-element subsets of A. Namely, since A has three elements, we know that there are three different 1-element subsets. In particular, we have $\{\text{Kobe}\}$, $\{\text{LeBron}\}$, and $\{\text{MJ}\}$ as our three 1-element subsets, so that all three of these subsets are elements of $P(A)$. We similarly have exactly three 2-element subsets of A. Namely, we have $\{\text{Kobe, LeBron}\}$, $\{\text{Kobe, MJ}\}$, and $\{\text{LeBron, MJ}\}$ as our three 2-element subsets. To see this, we can either explicitly check that there are no other distinct pairs of elements that we can choose from A, or we can notice that a 2-element subset of A must be missing precisely one element from A. And since we only have three choices of which element is missing from any given 2-element subset, we therefore have that there should be exactly three 2-element subsets. Finally, we have our 3-element subset which is the whole set $A = \{\text{Kobe, LeBron, MJ}\}$, since A has three elements. We therefore have that $P(A)$ has eight elements — one 0-element set (the empty set), three 1-element sets, three 2-element sets, and one 3-element set. Writing this out explicitly using our bracket notation gives

$$P(A) = \{\emptyset, \{\text{Kobe}\}, \{\text{LeBron}\}, \{\text{MJ}\}, \{\text{Kobe, LeBron}\}, \{\text{Kobe, MJ}\},$$
$$\{\text{LeBron, MJ}\}, \{\text{Kobe, LeBron, MJ}\}\}.$$

It is important to note the structure of the nested "{}" brackets, as these are what make explicit the fact that the **individual elements** of $P(A)$ are themselves sets.

3) Knowing that a set with N elements has a power set with 2^N elements, how many elements does the power set of the power set have? If we recall that $P(A)$ denotes the power set of A, then the question is to find the number of elements in $P(P(A))$. To start thinking about this problem correctly, first consider the set $A = \{a, b\}$ and write down all of the elements of $P(A)$. There should be four such elements. Then

$P(P(A))$ is the power set of a 4-element set. How many elements does this set have? Writing them all out explicitly is good practice.

Solution: Let us begin with the specific case when $A = \{a, b\}$. We then know that A has two elements and therefore that $P(A)$ has $2^2 = 4$ elements. Namely, the elements of $P(A)$ can be written explicitly as
$$P(A) = \{\emptyset, \{a\}, \{b\}, \{a, b\}\},$$
since we have one 0-element subset (the empty set), two 1-element subsets (the sets $\{a\}$ and $\{b\}$), and one 2-element subset (the whole set $\{a, b\}$). We now need to consider the power set of $P(A)$ itself, but this is not so bad because we are now simply taking the power set of a 4-element set. Thus, we know that the power set of $P(A)$, which is denoted[3] by $P(P(A))$, should have $2^4 = 16$ elements. Let us write these out explicitly.[4]

We first note that, as always, we have one 0-element subset of $P(A)$ and that is the empty set $\emptyset$. We then have four 1-element subsets of $P(A)$ and these are $\{\emptyset\}$, $\{a\}$, $\{b\}$, and $\{a, b\}$. It is **very** important to note the distinction between the 0-element subset $\emptyset$ of $P(A)$ and the 1-element subset of $P(A)$ given by $\{\emptyset\}$. Namely, the 0-element subset $\emptyset$ is the empty set — it is the subset of $P(A)$ with no elements at all in it. However, the 1-element subset $\{\emptyset\}$ of $P(A)$ is a set **whose single element** is the empty set. Namely, the empty set is **itself** an element of $P(A)$, so that we must form the 1-element subset $\{\emptyset\}$ of $P(A)$ if we want to consider $P(P(A))$. This may seem confusing and/or strange at first, but we should note that this should not be so. Namely, this is just another instance in which we must make the distinction between the set itself and a set that has this set as a

---

[3] This notation is used simply because we know that if A is the set that we want to consider the power set of, then we denote the power set by placing A in the parentheses of $P(\ )$. Thus, if $P(A)$ is the set whose power set we want to consider, then we denote this power set by placing $P(A)$ itself into the parentheses of $P(\ )$.

[4] The next two paragraphs are rather difficult. We present an easier example afterwards and it may be simpler to first look at this easier example and then return to the following two paragraphs. It may also be fruitful to first struggle through the next two paragraphs, then read the easier example, and then return again to the next two paragraphs.

single element. In particular, in order to consider $P(P(A))$, we must consider the empty set itself as being the empty subset of $P(A)$, **as well as** the set that has the empty set as an **element**. For example, the set $\{\emptyset\}$ is **not** the empty set, but rather a (non-empty) set whose single element is the empty set. Similarly, $\{\emptyset, \{b\}\}$ is a set with two elements, and each of these two elements are **subsets** of A. Namely, $\emptyset$ is a subset of A, as is $\{b\}$, and these two subsets make up the two **elements** of the set $\{\emptyset, \{b\}\}$.

Continuing on, we note that we have six 2-element subsets of $P(A)$ — we have $\{\emptyset, \{a\}\}$, $\{\emptyset, \{b\}\}$, $\{\emptyset, \{a,b\}\}$, $\{\{a\}, \{b\}\}$, $\{\{a\}, \{a,b\}\}$, and $\{\{b\}, \{a,b\}\}$. We note again the important use of nested brackets.[5] We then have four 3-element subsets, which can be seen by noting that for each 3-element subset we must exclude precisely one element from $P(A)$, and since $P(A)$ has four elements we have four choices of which element to exclude. Thus, our 3-element subsets of $P(A)$ are $\{\emptyset, \{a\}, \{b\}\}$, $\{\emptyset, \{a\}, \{a,b\}\}$, $\{\emptyset, \{b\}, \{a,b\}\}$, and $\{\{a\}, \{b\}, \{a,b\}\}$. Finally, we have one 4-element subset of $P(A)$, and that is $P(A)$ itself. Adding up our 0-, 1-, 2-, 3-, and 4-element subsets of $P(A)$, we see that we have $1+4+6+4+1$ subsets, which is precisely 16, as expected. Writing them all out and being very careful with our nested brackets, we have

$$P(P(A)) = \{\emptyset, \{\emptyset\}, \{a\}, \{b\}, \{a,b\}, \{\emptyset, \{a\}\},$$
$$\{\emptyset, \{b\}\}, \{\emptyset, \{a,b\}\}, \{\{a\}, \{b\}\}, \{\{a\}, \{a,b\}\}, \{\{b\}, \{a,b\}\},$$
$$\{\emptyset, \{a\}, \{b\}\}, \{\emptyset, \{a\}, \{a,b\}\}, \{\emptyset, \{b\}, \{a,b\}\},$$
$$\{\{a\}, \{b\}, \{a,b\}\}, \{\emptyset, \{a\}, \{b\}, \{a,b\}\}\}.$$

In order to not get lost in these nested brackets, let us consider $\{\emptyset, \{a\}, \{b\}\}$. This is a **single element** of the set $P(P(A))$ because it is a **subset** of $P(A)$. Namely, $\{\emptyset, \{a\}, \{b\}\}$ is the subset of $P(A)$ containing $\emptyset$, $\{a\}$, and $\{b\}$, which are all **elements** of $P(A)$ because they are all **subsets** of A.

Let us now look at an easier example of the power set of a 4-element set. Namely, we note that in the previous two paragraphs

---

[5] One should not be scared of expressions with large numbers of parentheses and/or brackets — we must simply and calmly work our way through the expressions keeping in mind that these parentheses and/or brackets are only there to organize the logical structure of the expression.

all we have done is explicitly write out the elements in the power set of a 4-element set, and the only difficulty arose due to subtleties regarding the empty set as well as with the abundance of nested brackets. Let us therefore now look at an example where neither of these difficulties are present. Namely, let us ask about the power set of the set $B = \{1, 2, 3, 4\}$, which is another perfectly good 4-element set.

We again know that the power set $P(B)$ of B contains 16 elements, since B has 4 elements and $2^4 = 16$. Indeed, we know that B has one 0-element subset (namely, the empty set $\emptyset$) and four 1-element subsets given by $\{1\}$, $\{2\}$, $\{3\}$, and $\{4\}$. Thus, so far this gives us five elements in $P(B)$. We also see that there are six 2-element subsets of B, given by $\{1, 2\}$, $\{1, 3\}$, $\{1, 4\}$, $\{2, 3\}$, $\{2, 4\}$, and $\{3, 4\}$. One way to see this is to begin by taking all the 2-element subsets that have the element 1 in them (of which there are three), then taking all the 2-element subsets that do **not** have the element 1 in them but **do** have the element 2 in them (of which there are 2), and finally taking all the 2-element subsets that do not have either the element 1 nor the element 2 in them but do have the element 3 in them (of which there is only 1). In this way we can be sure not to miss any 2-element subsets.

We now move on to the 3-element subsets of B, and we see that there are four of these. Namely, each 3-element subset of B corresponds to excluding one element from B, and since there are four elements of B that we can possibly leave out of any given 3-element subset, there are four different 3-element subsets. These are given by $\{1, 2, 3\}$, $\{1, 2, 4\}$, $\{1, 3, 4\}$, and $\{2, 3, 4\}$. Finally, we get one 4-element subset, and that is the set $\{1, 2, 3, 4\}$ itself. Adding up our 0-, 1-, 2-, 3-, and 4-element subsets of B, we see that there are indeed a total of 16 subsets and they can be explicitly written down as

$$P(B) = \{\emptyset, \{1\}, \{2\}, \{3\}, \{4\}, \{1, 2\}, \{1, 3\}, \{1, 4\}, \{2, 3\}, \{2, 4\}, \{3, 4\},$$
$$\{1, 2, 3\}, \{1, 2, 4\}, \{1, 3, 4\}, \{2, 3, 4\}, \{1, 2, 3, 4\}\}.$$

If we now return to our initial problem of explicitly writing down the power set $P(P(A))$ of the set $P(A) = \{\emptyset, \{a\}, \{b\}, \{a, b\}\}$, we see that we would obtain the same result if we first do the analysis that we just performed on the set $B = \{1, 2, 3, 4\}$ and then replace the

SOLUTIONS TO CHAPTER 5

element 1 in B with the element $\emptyset$ in $P(A)$, as well as replace the element 2 in B with the element $\{a\}$ in $P(A)$, the element 3 in B with the element $\{b\}$ in $P(A)$, and the element 4 in B with the element $\{a,b\}$ in $P(A)$. Indeed, by writing $\emptyset$ wherever we see 1 above, $\{a\}$ wherever we see 2, $\{b\}$ wherever we see 3, and $\{a,b\}$ wherever we see 4, then we recover the long expression for $P(P(A))$ that we found before, where even the nested brackets work out in the correct way! This is not surprising, however, since taking the power set of any 4-element set will involve the exact same steps, regardless of what the elements of that set really are. In our second example, the elements of our 4-element set are the numbers 1, 2, 3, and 4, whereas in our previous example the elements were subsets of the set $A = \{a,b\}$. The beauty and power of set theory is that, up to a renaming of the elements, our analysis is independent of whether we are dealing with a set of numbers or a set of subsets or a set of NBA players.

To recap, we have found that the power set of $A = \{a,b\}$ has four elements, and that the power set of this power set has 16 elements. This is all in line with the statement that the power set of a set with N elements has 2^N elements. Namely, for A, $N = 2$ and we know that $2^2 = 4$, which is indeed the number of elements in $P(A)$. Then, since $P(A)$ has 4 elements, the "N" for $P(A)$ is 4. We then note that $2^4 = 16$ which is indeed in line with our finding that $P(P(A))$ has 16 elements. Let us now turn to the more general case where we **do not** specify the number of elements in A.

Namely, let us now suppose that A has N elements. We then know that $P(A)$ has 2^N elements. In order to see how many elements the power set of $P(A)$ has, we simply apply this logic again to the set $P(A)$. Namely, we can let $M = 2^N$, and then we are just asking what the number of elements in the power set of a set with M elements is. We already know this answer — such a set has 2^M elements. Now, recalling that $M = 2^N$, we see that $P(P(A))$ has 2^{2^N} elements. For example, if A has 2 elements, then $P(P(A))$ has $2^{2^2} = 2^4 = 16$ elements (as we have seen above), if A has 3 elements, then $P(P(A))$ has $2^{2^3} = 2^8 = 256$ elements, and if A has 4 elements, then $P(P(A))$ has $2^{2^4} = 2^{16} = 65,536$ elements. Clearly, the size of $P(P(A))$ grows extremely fast as the size of A grows!

Solutions To Chapter 5

1) Let $A = \{a, b, c\}$ and $W = \{14, 57, 85, 0\}$. **State whether or not each of the following are functions from A to W.**

(a) **The assignment that is defined by** $f(a) = 57$ **and** $f(b) = 0$.

(b) **The assignment that is defined by** $f(a) = b$, $f(b) = 85$, **and** $f(c) = 14$.

(c) **The assignment that is defined by** $f(a) = 57$, $f(b) = 57$, **and** $f(c) = 85$.

(d) **The assignment that is defined by** $f(a) = 0$, $f(b) = 57$, $f(a) = 14$, **and** $f(c) = 57$.

Solution: (a) This assignment does **not** specify a valid function. The reason for this is that a function must send **every** element in its domain (in this case, the domain is $A = \{a, b, c\}$) to an element in its codomain (in this case, the codomain is $W = \{14, 57, 85, 0\}$), and moreover it must send every element in its domain to **only one** element in its codomain. For the assignment given in this part, the element c in the domain is never sent anywhere, and therefore this assignment does not define a valid function.

(b) This assignment also does **not** specify a valid function. This is because the element a in the domain is not sent to an element in the codomain, as it should be. Namely, a is sent to b and b is not in the codomain W.

(c) This assignment **does** specify a valid function. Namely, we can see by inspection that each element in A gets sent to some element in W, and moreover that each element in A gets sent to **only** one element in W. Since these are the only requirements for the assignment to be considered a valid function, we see that this f is indeed a valid function.

(d) This assignment does **not** specify a valid function. The reason for this is that the element a in the domain gets sent to **two different** elements in the codomain W. Namely, the element a gets sent to both 0 and 14 in the codomain, and this is not allowed based on our definition of a function. Thus, this f is not a valid function.

SOLUTIONS TO CHAPTER 5

2) Let $\text{Num}_1 = \{1,2,3\}$ and $\text{Num}_2 = \{4,5,6\}$. Define three different functions from Num_1 to Num_2. Namely, come up with a function f such that $f : \text{Num}_1 \to \text{Num}_2$, and then do it two more times (coming up with different functions). Then define three different functions from Num_2 to Num_1. For each function that you define, consider drawing the corresponding "arrow diagram," analogous to Figure 5.1.

Solution: There are many more than three possible functions from $\text{Num}_1 = \{1,2,3\}$ to $\text{Num}_2 = \{4,5,6\}$, but we will only present three here. We may not all choose the same three functions to define, but hopefully the three that we present here will be sufficient for us to check the correctness (or lack thereof) of the other possibilities on our own. Let us define our first function to send 1 in Num_1 to 4 in Num_2, 2 in Num_1 to 5 in Num_2, and 3 in Num_1 to 6 in Num_2. If we denote this function by f_1, then we have **defined** this function to be such that $f_1(1) = 4$, $f_1(2) = 5$, and $f_1(3) = 6$. The arrow diagram for this function[6] is shown in Figure 21.2. We recall that a function

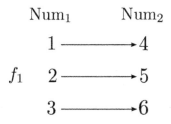

Figure 21.2: Visual representation of the function f_1.

must send **each** of the elements in its domain (which, in this case, is Num_1) to some element in its codomain (which, in this case, is

---

[6] Let us not be confused by expressions like $f_1(1) = 4$. We recall from the main text of this chapter that $f_1(1) = 4$ simply means that "the function f_1 sends the element 1 to the element 4." Thus in particular, the subscript 1 and the 1 in the parentheses are completely different. The subscript is just one possible way of denoting (i.e., labeling) our function, in that we have chosen to denote our function by f_1 instead of by f or g or h. The 1 in the parentheses, however, is the element of Num_1 that is under consideration and that is being sent to Num_2 by the function f_1.

Num_2), and it must send each element in the domain to **only** one element in the codomain. We immediately see that f_1 does indeed satisfy these two requirements, and therefore it is a valid function.

Let us now define a second function, denoted by f_2, from Num_1 to Num_2, so that[7] $f_2 : Num_1 \to Num_2$. In particular, let us define f_2 to be the function that sends all of the elements in Num_1 to the element 5 in Num_2. We note that this does indeed satisfy both requirements of a function,[8] and is therefore a valid function. The arrow diagram for this function is shown in Figure 21.3.

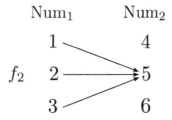

Figure 21.3: Visual representation of the function f_2.

As a third example of a function from Num_1 to Num_2, let us define the function f_3 to send both 1 and 3 in Num_1 to 4 in Num_2, and to send 2 in Num_1 to 6 in Num_2. In symbols, we define f_3 to be such that $f_3(1) = 4$, $f_3(2) = 6$, and $f_3(3) = 4$. We again see that these assignments satisfy the requirements of a function, as it must. The arrow diagram for this function is shown in Figure 21.4.

We now need to define three functions going the other way, namely **from** Num_2 to Num_1. Let us call the first function g_1 so that $g_1 : Num_2 \to Num_1$, and let us define g_1 by the assignments $g_1(4) = 1$, $g_1(5) = 2$, and $g_1(6) = 3$. We see that this is a valid function (in the sense that it satisfies the two requirements of a function), and the arrow diagram of this function is shown in Figure 21.5.

As a second example of a function from Num_2 to Num_1, we can define $g_2 : Num_2 \to Num_1$ by the assignments $g_2(4) = 3$, $g_2(5) = 2$

---

[7]Note that since we read "$f_2 : Num_1 \to Num_2$" as "f_2 is a function from Num_1 to Num_2," this sentence is indeed grammatically correct, just as math should be.

[8]Namely, that each element in the domain (which is Num_1 here) is sent to some element in the codomain (which is Num_2 here), and that each element in the domain is sent to **only** one element in the codomain.

SOLUTIONS TO CHAPTER 5   249

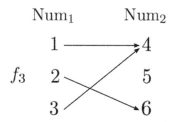

Figure 21.4: Visual representation of the function f_3.

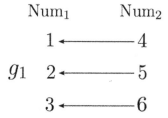

Figure 21.5: Visual representation of the function g_1.

and $g_2(6) = 1$. We can quickly check that these assignments do define a valid function, and the arrow diagram of this function is shown in Figure 21.6.

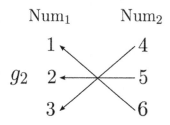

Figure 21.6: Visual representation of the function g_2.

Finally, as a third example of a function from Num_2 to Num_1, we can define $g_3 : \text{Num}_2 \to \text{Num}_1$ to send all of the elements in Num_2 to the element 1 in Num_1. Namely, we can define g_3 to be such that $g_3(4) = 1$, $g_3(5) = 1$, and $g_3(6) = 1$. The arrow diagram for this function is shown in Figure 21.7.

There are, of course, many other possible functions from Num_2

to Num_1, however we will not show them all here. We hope that the examples that we gave here are sufficiently illuminating so as to allow ourselves to be able to check on our own whether or not other possible functions are valid.

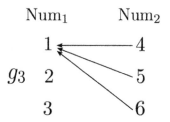

Figure 21.7: Visual representation of the function g_3.

3) Let $C = \{\text{Mt. Everest}, \text{cheeseburger}, 4\}$. **Define two different functions from C to itself. Namely, come up with a function f such that $f : C \to C$, and then do it again by defining a different function. Again, consider drawing the "arrow diagrams" for the functions you define.**

Solution: In principle, this problem is no different from the previous problem. The only difference is that now the domain and the codomain are the same — namely, they are both C — but there is nothing wrong with this. In particular, there is nothing in the definition of a function that prohibits us from defining a function from a set to itself, and indeed it is often extremely interesting and fruitful to study the kinds of functions one can define from a certain kind of mathematical structure back to itself. Here, we will provide just two examples (as the problem asks for) of functions from C to itself, though there are indeed many more possibilities that we could have considered.

The first example will be the function that sends every element in C back to itself. Namely, if we call this function f_1, then we have just defined $f_1 : C \to C$ to be such that $f_1(\text{Mt. Everest}) = \text{Mt. Everest}$, $f_1(\text{cheeseburger}) = \text{cheeseburger}$, and $f_1(4) = 4$. We can quickly see that this is indeed a valid function, since each element in the domain (in this case, the domain is C) gets sent to one and only one element in the codomain (in this case, the codomain is also C). The arrow diagram of this function is shown in Figure 21.8.

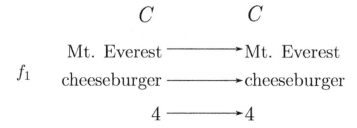

Figure 21.8: Visual representation of the function f_1.

As a second example of a function from C to itself, we can send all of the elements in C to the same element in C. For example, we can send all of the elements in C to the element "cheeseburger" in C. If we call this function f_2, then we would have f_2(Mt. Everest) = cheeseburger, f_2(cheeseburger) = cheeseburger, and $f_2(4)$ = cheeseburger. We can quickly see that this is indeed a valid function, and the arrow diagram for this function is shown in Figure 21.9.

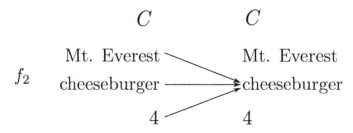

Figure 21.9: Visual representation of the function f_2.

4) (a) Let $A = \{a, b\}$ and let $X = \{0, 1\}$. **How many possible functions are there from A to X?**

(b) Let $A = \{a, b, c\}$ and let $X = \{0, 1\}$. **How many possible functions are there from A to X?**

Solution: (a) In the previous two exercises we have only been asked to explicitly define a few of the possible functions between two given sets. Now, however, we are asked to enumerate **all possible** functions from the set $A = \{a, b\}$ to the set $X = \{0, 1\}$. We must therefore find a systematic way to move through all of the possibilities, so as to ensure that we do not miss any.

We begin by noting that any function from A to X must send a in A to some element in X, and it must also send b in A to some element in X. Let us therefore begin with the element a. Since a must be sent to either 0 or 1, let us begin by considering all of the functions that involve sending a to 0. We immediately see that there are two such functions — either a goes to 0 and b goes to 0, or a goes to 0 and b goes to 1. Thus, we have found that there are two possible functions from A to X such that a goes to 0. Let us now consider the functions from A to X that send a to 1. We again see that there are two such functions — either a goes to 1 and b goes to 0, or a goes to 1 and b goes to 1. Thus, there are two functions from A to X such that a is sent to 1.

Since we have found that there are two functions that involve a going to 0 and two functions that involve a going to 1, we see that we have at least four possible functions from A to X. However, since **any** function from A to X must send a to either 0 or 1, we see that we have already covered all of the possible functions. Thus, there are a **total** of four functions from A to X — one that sends both a and b to 0, one that sends a to 0 and b to 1, one that sends a to 1 and b to 0, and one that sends both a and b to 1.

Another way to see that there are four possible functions from A to X is to think of things in the following way. We have two choices of where to send a — we can send a either to 0 or to 1. For each of these choices, we have two choices of where to send b — we can send b either to 0 or to 1. Thus, for each of the two choices of where to send a, we get two choices of where to send b, and we therefore have $2 \times 2 = 4$ choices total. Since each pair of choices (of where to send a and where to send b) corresponds to a different function from A to X, we see that we have four possible functions in total. This way of looking at things will indeed be the best way to think of things when we solve the next part of this exercise below.

(b) We now want to answer the same question that we did in the previous part, only where $A = \{a, b, c\}$. Let us first use the logic that

SOLUTIONS TO CHAPTER 5   253

we used at the end of the solution to the first part. Namely, we see that we have two choices of where to send a — either to 0 or to 1. For each of these two choices, we have two choices of where to send b — either to 0 or to 1. This makes for four possible choices in total of where to send a and b. And for each of these four choices, we have two choices of where to send c — either to 0 or to 1. Thus, since we have four possible choices of where to send a and b (which we already know from the previous part of this exercise, which showed that there are four possible functions from $\{a, b\}$ to $\{0, 1\}$), and since for each of these choices we have two possible choices of where to send c, we see that we have eight possible choices of where to send all of the elements $\{a, b, c\}$. Since each of these choices corresponds to a different function, we see that there are eight possible functions from $\{a, b, c\}$ to $\{0, 1\}$.

Indeed, we can extend this logic to any set with a finite number of elements. Namely, let us suppose that[9] $A = \{1, 2, 3, ..., N\}$ so that, in particular, the set A has N elements. Suppose also that we want to know how many functions there are from A to $X = \{0, 1\}$. The answer is that there are 2^N such functions, and we will see why momentarily. First, let us note that this is in line with the two results that we have proven so far. Namely, when A has two elements (as it does in the first part of this exercise), then $N = 2$ and our claim says that there should be $2^2 = 4$ possible functions from A to X. This is precisely what we proved above. Similarly, when A has three elements (as it does in the previous paragraph), then $N = 3$ and our claim says that there should be $2^3 = 8$ possible functions from A to X. This is also precisely what we proved in the previous paragraph.[10]

---

[9]In order to see what the following notation really means, it is best to look at a few examples. Namely, if $N = 5$ then $A = \{1, 2, 3, 4, 5\}$, if $N = 2$ then $A = \{1, 2\}$, if $N = 1$ then $A = \{1\}$, and if $N = 7$ then $A = \{1, 2, 3, 4, 5, 6, 7\}$. Here, we are just letting N be anything and considering the situation for any N.

[10]We note that the present case with $N = 2$ gives the set $A = \{1, 2\}$, and that the number of functions from $\{1, 2\}$ to $\{0, 1\}$ is the same as the number of functions from $\{a, b\}$ — which is the set considered earlier — to $\{0, 1\}$. Similarly, for $N = 3$, we have $A = \{1, 2, 3\}$ in the present case, and the number of functions from $\{1, 2, 3\}$ to $\{0, 1\}$ is the same as the number of functions from $\{a, b, c\}$ to $\{0, 1\}$. All that matters when considering the number of functions from one set to another is the number of elements in each set, and not the precise details of what those elements are.

To see that there are 2^N possible functions from A to X whenever A has N elements, we simply need to extend the logic that we have already used twice in this solution. Namely, we have N elements that we need to send to either 0 or 1. This means that for each element, we have two choices of where it is sent to. In other words, when $A = \{1, 2, ..., N\}$, we have two choices of where to send 1 (either to 0 or 1) and two choices of where to send 2 (either to 0 or 1). This gives a total of 2×2 choices of where to send 1 and 2. For each of these choices we have two choices of where to send 3 (either to 0 or 1), and so this gives a total of $2 \times 2 \times 2$ choices of where to send 1, 2, and 3. For each of these choices, we have two choices of where to send 4 (either to 0 or 1), and so this gives a total of $2 \times 2 \times 2 \times 2$ choices of where to send 1, 2, 3, and 4. Continuing this on until we send all of our N elements to either 0 or 1, we see that we have $2 \times 2 \times ... \times 2$ choices that we can make, where the "..." simply means that in total there are N factors of 2 (since we do not know what N is explicitly, we have to use this "..." notation). However, since $2 \times 2 \times ... \times 2$ simply denotes the fact that we are multiplying 2 to itself N times, we have that there are precisely 2^N choices that we can make.[11] Since each possible choice corresponds to a different possible function, we have found that there are 2^N possible functions from any set with N elements to the set $X = \{0, 1\}$. We also note that we can generalize X to **any** set with two elements — the fact that these two elements are 0 and 1 (as opposed to, say, Kobe and LeBron) is of no importance for what we have done here. This result is what helps us prove that a set with N elements has 2^N possible subsets, and this proof is given in Appendix B.

Returning to the particular case when $A = \{a, b, c\}$ as given in the problem, let us explicitly state the eight possible functions so that we can see that our logic actually works via an example. Let us organize things according to the number of elements in A that get sent to 1 in X. Namely, a function from A to X can send 0, 1, 2, or 3 of the elements in A to the element 1 in X, and we will use this to organize our analysis.

There is only one function from $A = \{a, b, c\}$ to $X = \{0, 1\}$ that

---

[11] This is simply because 2^N also denotes the process of multiplying 2 to itself N times.

sends zero elements to 1, and that is the function that sends a, b, and c all to 0. There are then three functions from A to X that send only one element in A to 1 in X. These three functions are the function that sends a to 1 and both b and c to 0, as well as the function that sends b to 1 and both a and c to 0, as well as the function that sends c to 1 and both a and b to 0. There are now three functions that send two elements in A to 1 in X. These three functions are the function that sends a to 0 and both b and c to 1, the function that sends b to 0 and both a and c to 1, and the function that sends c to 0 and both a and b to 1. Finally, there is only one function that sends three elements in A to 1 in X, and that is the function that sends a, b, and c all to 1 in X. Thus, we have $1 + 3 + 3 + 1 = 8$ possible functions, in agreement with our previous analysis.

Solutions To Chapter 6

1) Define the set $A = \{\text{cow}, 4\}$ and the set $B = \{\text{Magic}, \text{Bird}, \text{Mamba}\}$. Are there any injective functions from A to B? Are there any surjective functions from A to B? Are there any bijective functions from A to B? For each question that has a positive answer, find out how many there are. I.e., if there are injective functions from A to B, find out how many different possible injective functions there are (this is a hard, but possible question to answer — see the solutions after giving it some thought).

Solution: Let us begin by answering the question of whether or not these various kinds of functions even exist before we start enumerating them. Namely, let us first ask whether or not there exists an injective function from A to B. Indeed, the answer here is yes, since A has fewer elements than B. Recall that we can assign elements in A to elements in B in an injective way only if the number of elements in A is less than or equal to the number of elements in B. We should think again about the example of the school dance and assigning boys to girls — if there are more boys than girls, then if we define a function from the boys to the girls there will end up being some girls who have more than one dance partner simply because the definition of a function requires us to send each boy to some girl. However, if the number of boys is less than or equal to the number of girls, then it is possible to assign boys to girls in an injective way. Namely, it is possible to assign boys to girls in such a way that no girl is dancing with more than one boy.

For similar reasons, there does **not** exist a surjective function from A to B, simply because A has fewer elements than B and so we cannot "hit" all of the elements in B with the elements in A. This follows from the fact that, by definition of a function, we cannot send any elements in A to more than one element in B. Thus, since we cannot "kill two birds" (i.e. hit two elements) in B with only one stone (i.e. element) from A, and since A has fewer stones (i.e. elements) than B has birds (i.e. elements), there will always be some birds left over in B and therefore any function from A to B will not be surjective. Additionally, since there cannot exist a surjective function from A to B, there also cannot exist a bijective function from A to B, simply because a bijective function from A to B would also be a surjective function from A to B (simply because a bijective function is, by definition, surjective and injective), and we just showed that there can be no surjective functions from A to B. There are of course functions from A to B that are **neither** injective nor surjective (and thus not bijective either), but there at least **exists** functions from A to B that are injective. Let us now turn to counting how many such functions there are.

We begin by noting that we **do not** want the total number of functions from A to B — we only want the total number of **injective** functions. An injective function from A to B will, just like any other function, send "cow" in A to some element in B. This gives us three choices for where to send "cow," since we could send "cow" to "Magic," "Bird," or "Mamba" in B. Now, if we were interested in the total number of functions (without the restriction to injective functions), then we could also say that we have three choices for where to send 4 in A, since in this case we would also be able to send 4 to any of the three elements in B. This would give a total of 9 functions (not necessarily injective), since each of the three choices for where to send "cow" is accompanied by three choices for where to send 4, thus giving a total of $3 \times 3 = 9$ choices, each of which corresponds to a different function.

However, since we are restricting ourselves to **injective** functions from A to B, we **cannot** send 4 to any element in B once we have already sent "cow" somewhere. This is because we must make sure that we do not send 4 to the same location that we sent "cow," since if we did this then we will have sent 4 and "cow" to the same element in B and our function will not be injective. Thus, once we have made one of our three choices for where to send "cow," we only have **two** choices left for where to send 4 — namely, we can only send 4 to one of the two elements in B that was **not** hit by "cow." For example, if we send "cow" to "Mamba," then we can only send 4 to either "Magic" or "Bird" if we want this function to be injective. Thus, since each of the three choices for where to send "cow" is accompanied by **only two** choices for where to send 4, we see that we have a total of $3 \times 2 = 6$ choices for where to send "cow" and 4, which means that there are a total of six injective functions from A to B.

If we recall that we found above that there should be nine functions **in total** (i.e., not necessarily injective) from A to B, then our result that there are six injective functions from A to B tells us that only three of the general functions from A to B are **not** injective. Let us explicitly check all of this by constructing all nine of the possible functions from A to B, beginning with the six injective ones and then filling in the last three non-injective ones. We will list these all now, and we will do so quickly because we have seen this sort of thing many times by now.

The first injective function that we can consider is the one that sends "cow" to "Magic" and 4 to "Bird." The second injective function that we can consider is the one that sends "cow" to "Magic" and 4 to "Mamba." The third injective function that we can consider is the one that sends "cow" to "Bird" and 4 to "Magic." The fourth injective function that we can consider is the one that sends "cow" to "Bird" and 4 to "Mamba." The fifth injective function that we can consider is the one that sends "cow" to "Mamba" and 4 to "Magic." The sixth injective function that we can consider is the one that sends "cow" to "Mamba" and 4 to "Bird."

Now, the only way for a function whose domain only has two elements to be non-injective is for both of those elements to be mapped to the same element in the codomain. This is because if these two elements were mapped to **different** elements in the codomain, then the function would be injective! Thus, we get three non-injective functions — the one that sends both "cow" and 4 to "Magic," the one that sends both "cow" and 4 to "Bird," and the one that sends both "cow" and 4 to "Mamba." Adding these three non-injective functions to the six injective functions that we described in the previous paragraph indeed gives us a total of nine functions, as expected.

2) Define two different surjective functions from the set $A = \{1, 2, a, b\}$ to the set $B = \{\text{cow}, \text{horse}\}$.

Solution: There are indeed many more than two surjective function from A to B in this problem, and we will give as explicit examples only four of them. Therefore, any two of the following four would suffice as a solution, as would any two valid surjective functions that we do not explicitly mention here — all that matters is that **both** "cow" and "horse" have some element in A that is sent to them via our function, as this is the definition of surjectivity.

Let us therefore construct our surjective functions by first making sure both "cow" and "horse" are hit and then sending our remaining elements in A anywhere we would like. In particular, let us define our first function f_1 to send 1 in A to "cow" in B, and 2 in A to "horse" in B. In other words, we have defined f_1 so that $f_1(1) = $ cow and $f_1(2) = $ horse. By doing this we have guaranteed that our function is surjective, since "cow" and "horse" are now both already hit. However, f_1 is not yet a function since we have not sent a or b anywhere. Since f_1 has already hit both elements in B, however, it does not really matter where we send a and b so long as we send them somewhere. Let us therefore define f_1 to send a and b both to "cow," so that we have $f_1(a) = $ cow and $f_1(b) = $ cow. Thus, f_1 sends every element in A to "cow" except for the element 2, since 2 is sent to "horse" in order to make our function surjective.

We can indeed be done with this problem already by defining a second function f_2 to behave almost exactly like f_1, where the only difference lies in where f_2 sends a and/or b. For example, let us define f_2 to send 1 to "cow" and 2 to "horse" (just like f_1 does), and let us also define f_2 to send a to "cow" as well (just like f_1 does). Then, in order to ensure that f_2 and f_1 are not the exact same function, let us define f_2 to send b to "horse" (whereas f_1 sends b to "cow"). We again have that both "cow" and "horse" are hit by f_2 so that f_2 is surjective, and since f_2 is not the exact same function as f_1 we have successfully defined two **distinct** surjective functions from A to B.

We can define a third surjective function f_3 from A to B by using the same method — let f_3 agree with f_1 and f_2 on where it sends 1 and 2, and only differ in where it sends a and/or b. In particular, if we define f_3 to send 1 to "cow" and 2 to "horse" (just as we did f_1 and f_2), then we know that f_3 will be surjective once we make it an actual function by sending a and b somewhere. Let us choose to send both a and b to "horse" with f_3, since neither f_1 nor f_2 does this. Thus, f_3 sends every element in A to "horse" except for the element 1 in A, which it sends to "cow." We already know that this is a surjective function and since it is not identical to either f_1 or f_2, we see that we have successfully defined a third surjective function from A to B.

We promised we would explicitly define four surjective functions from A to B (even though the problem only asks for two), and we could indeed define a fourth function that agrees with f_1, f_2, and f_3 on the elements 1 and 2 (namely, by sending 1 to "cow," 2 to "horse," a to "horse," and b to "cow"). However, let us define our fourth function f_4 in such a way that it does **not** agree with f_1, f_2, or f_3 on the elements 1 and 2. Namely, let us define f_4 to send both 1 and 2 to the element "cow." Our function f_4 is then not surjective unless we send either a or b (or both) to "horse" so that we have indeed hit all of the elements in B. Any of the possible choices of how to send a and b to B with one (or both) of them hitting "horse" is valid, and let us arbitrarily define f_4 to send a to "cow" and b to "horse." We then have that f_4 is surjective since both "cow" and "horse" are hit by f_4. We also have that f_4 is not identical to any of the functions f_1, f_2, or f_3 that we have defined before since all three of those functions send 1 and 2 to different elements, whereas f_4 sends 1 and 2 to the same element (namely, "cow"). We therefore have successfully defined a fourth surjective function from A to B, as promised. There are of course **many** other possible surjective functions from A to B, and hopefully these examples have given us enough to be able to check the validity of other possibilities that we did not explicitly address here.

3) How many different bijective functions are there from the set $A = \{1, 2, 3\}$ to the set $B = \{a, b, c\}$?

Solution: There are (at least) two ways to solve this problem, and we will present both here. The first way that we will present will be the quicker, more elegant way, but it will have the downside of not telling us explicitly **what** the possible bijective functions are. This does not matter for this problem in particular, since the problem only asks for **how many** bijective functions there are from A to B and not **what** these functions are, but it is indeed enlightening to see some examples of these bijective functions. To remedy this, our second method will be less elegant but will have the benefit of showing us exactly what the bijective functions actually "look like." In particular, for our second method we will simply and explicitly define all of the possible bijective functions.

For the quicker method, we have to recall that we are looking only for the **bijective** functions from A to B. Keeping this in mind, we can begin to use the same logic that we have used before in similar problems. Namely, as with any function from A to B, we must send 1 to something in B. This gives us three choices for where to send 1, since B has three elements. Now, for a **general** function from A to B, for each choice of where to send 1 we would also have three choices for where to send 2, and for each of these choices we would have three choices for where to send 3. This would give us a total of $3 \times 3 \times 3 = 27$ possible choices of where to send 1, 2, and 3, and therefore we would know that there are 27 possible **general** functions from A to B.

However, things are not so simple for bijective functions. This is because once we make one of our three choices for where to send 1, we **cannot** send 2 to this element, for otherwise 1 and 2 would be sent to the same element and our function would not be injective, and therefore our function would not be bijective. Thus, for each of the three choices for where to send 1, we only have **two** choices for where to send 2 (namely, we can send 2 to either of the two elements

SOLUTIONS TO CHAPTER 6

in B that were **not** hit by 1). This gives us $3 \times 2 = 6$ total choices for where to send 1 and 2. Now, in order to figure out where to send 3 while still ensuring that our function is bijective, we must remember that we cannot send 3 to **either** of the elements that were hit by 1 or 2. Since B has three elements and since two of these elements have already been hit by 1 and 2, we are left with only one choice for where to send 3! In other words, once we make our choices for where to send 1 and 2, we **no longer have a choice** for where to send 3, assuming we want our function to be bijective. Thus, we only have six real choices to make. Since each such choice corresponds to a different bijective function from A to B, we see that we have a total of six bijective functions from A to B.

To check all of this, let us actually construct all the possible bijective functions from A to B and ensure that there are indeed only six of them. Since our bijective functions must all send 1 to something in B, let us first consider the bijective functions that send 1 to a. Once we have sent 1 to a, then we can send 2 only to either b or c. Suppose we send 2 to b. Then we **must** send 3 to c in order for our function to be bijective. Thus, our first bijective function (which we will choose to denote by f_1) is defined by $f_1(1) = a$, $f_1(2) = b$, and $f_1(3) = c$. If instead of sending 2 to b we had chosen to send 2 to c, then we would have been **forced** to send 3 to b (assuming we have already sent 1 to a). Thus, our second bijective function (which we will choose to denote by f_2) is defined by $f_2(1) = a$, $f_2(2) = c$, and $f_2(3) = b$. This exhausts all of the possibilities for a bijective function from A to B that sends 1 to a.

Let us therefore move on to the possible bijective functions from A to B that send 1 to b. We can then send either 2 to a and 3 to c, or vice versa — namely, send 2 to c and 3 to a. Denoting these two functions respectively by f_3 and f_4, we have that f_3 is defined by $f_3(1) = b$, $f_3(2) = a$, and $f_3(3) = c$, and that f_4 is defined by $f_4(1) = b$, $f_4(2) = c$, and $f_4(3) = a$. These exhaust the possibilities for a bijective function from A to B that sends 1 to b, and so far the count for the total number of bijective functions from A to B is four.

We can now move on to the possible bijective functions from A to B that send 1 to c. Once we send 1 to c, our only two options are to send 2 to a and 3 to b or vice versa — namely, to send 2 to b

and 3 to a. Denoting these two functions respectively by f_5 and f_6, we have that f_5 is defined by $f_5(1) = c$, $f_5(2) = a$, and $f_5(3) = b$, and that f_6 is defined by $f_6(1) = c$, $f_6(2) = b$, and $f_6(3) = a$. This exhausts all of the possibilities for a bijective function from A to B that sends 1 to c. Since any bijective function (indeed, any function at all) from A to B will send 1 to either a, b, or c, and since we have exhausted the possibilities for bijective functions in each case, we see that the functions f_1, f_2, f_3, f_4, f_5, and f_6 are all of the possible bijective functions from A to B. Thus, by explicitly constructing all possible bijective functions from A to B, we see that there are exactly six possibilities, which is exactly in line with what we found above using our more elegant method of simply counting choices. Thus, we can be very sure in saying that the answer to this problem is six.

Solutions To Chapter 10

1) **Prove that "infinity type 1" divided by two is still "infinity type 1." Do this by realizing that the set of even numbers is a meaningful way of describing "infinity type 1 divided by two," since we can evenly split $\mathbb{N} = \{1, 2, 3, ...\}$ into odd and even parts. Thus, in a clear sense, the even numbers alone (as well as the odd numbers alone) have "half as many" elements as $\mathbb{N}$. Then define a bijective function from $\mathbb{N} = \{1, 2, 3, ...\}$ to $B = \{2, 4, 6, 8, ...\}$, thus showing that $\mathbb{N}$ and B actually do have the same cardinality!**

Solution: We do just what the hint suggests. Namely, let the set B be as the hint describes, so that $B = \{2, 4, 6, 8, ...\}$ is a meaningful notion of "infinity type 1 divided by 2." We now want to define a bijective function from $\mathbb{N}$ to B to show that these two sets have the same cardinality, and we will do so simply by sending each element in $\mathbb{N}$ to "2 times itself." Namely, we send 1 in $\mathbb{N}$ to 2 in B, 2 in $\mathbb{N}$ to 4 in B, 3 in $\mathbb{N}$ to 6 in B, 4 in $\mathbb{N}$ to 8 in B, and so on, forever. This is clearly injective, since no two distinct elements in $\mathbb{N}$ are sent to the same element in B. To see this, we note that if two elements in $\mathbb{N}$ are sent to the same element in B, then that means that one of these two elements multiplied by 2 is equal to the other element multiplied by 2. But if "something" multiplied by 2 is equal to "something else" multiplied by 2, then the "something" must be equal to the "something else." Therefore, the only way two elements in ℕ can be sent to the same element in B by our function is if these two elements in ℕ are actually the **same** element, and this is nothing but the definition of injectivity for our function.

Moreover, this function (which sends each element in ℕ to the element in B which is itself multiplied by 2) is surjective since each element N in B is hit precisely by "N divided by 2" in ℕ. For example, the element 26 in B is hit by 13 in ℕ, the element 100 in B is hit by 50 in ℕ, and the element 168 in B is hit by 84 in ℕ. Since all of the elements in B are even, we can always divide them by two to find the element in ℕ that maps to it. Thus this function is bijective (since we have just seen that it is injective and surjective), and so we have shown that "infinity type 1 divided by 2" is actually equal to "infinity type 1"!

As an extra challenge (for which we will not provide the solution), how would one adapt this argument to show that "infinity type 1" divided by 3 is again "infinity type 1"? What about "infinity type 1" divided by 4? What about "infinity type 1" divided by m for any finite positive whole number m? The only hint that we will give is that in the argument given above, we modeled "infinity type 1" divided by 2 by the set of numbers that are multiples of 2 (i.e., the even numbers), and then defined a bijective function from ℕ to this set by sending each number in ℕ to "two times itself." How does this change for these other cases?

Solutions To Chapter 16

1) How many elements are in the union of the sets $\{1, 2, 3, 4, 7\}$ and $\{3, 4, 5\}$? How many are in their intersection?

Solution: The union of the two sets $\{1, 2, 3, 4, 7\}$ and $\{3, 4, 5\}$ is the set whose elements are either in $\{1, 2, 3, 4, 7\}$ or in $\{3, 4, 5\}$. Thus, $\{1, 2, 3, 4, 7\} \cup \{3, 4, 5\} = \{1, 2, 3, 4, 5, 7\}$ and so there are six elements in the union. It is important to note that we do **not** need to "double count" the elements 3 and 4, as we might be tempted to since they are in both sets $\{1, 2, 3, 4, 7\}$ and $\{3, 4, 5\}$. This is simply because when we take the union of two sets we do not distinguish between identical elements — we simply include all of the elements that are in either of the two sets.

To answer the second part of the question, we must recall that the intersection of two sets is, by definition, the set whose elements are in **both** of the two sets in question. Thus, we have that the intersection of $\{1, 2, 3, 4, 7\}$ and $\{3, 4, 5\}$ is simply $\{3, 4\}$, since only 3 and 4 are in **both** of our initial sets. In symbols, this can be expressed as $\{1, 2, 3, 4, 7\} \cap \{3, 4, 5\} = \{3, 4\}$, and we see that the intersection has only two elements.

2) What is the union of the sets

$$\{p \in \mathbb{N} \mid p \text{ is even}\}$$

and

$$\{p \in \mathbb{N} \mid p \text{ is a multiple of 4}\}?$$

What is their intersection?

Solution: Let us let $A = \{p \in \mathbb{N} \mid p \text{ is even}\}$ and $B = \{p \in \mathbb{N} \mid p \text{ is a multiple of 4}\}$, and let us also note that B is a subset of A. This is because any multiple of 4 (like 4, 8, 12, 16, 20, etc.) is also an even number, and since B is the set of all elements in $\mathbb{N}$ which are multiples of 4, we see that every element that is in B is also in A. Thus, the union of these two sets is simply A because B does not "bring anything new to the table," as all of its elements are already in A. Similarly, the intersection of A and B is simply B, because any element in B is in both A and B, which is the definition of the intersection of two sets.

This is indeed a general phenomenon. Namely, if B is a subset of A then $A \cup B = A$ and $A \cap B = B$. This is precisely the answer to Exercise 4 in this chapter, so for a more detailed discussion as to **why** this is true we should take a look below at the solution to that problem.

SOLUTIONS TO CHAPTER 16

3) What is the intersection of the sets $\{a \in \mathbb{N} \mid a \text{ is odd}\}$ and $\{a \in \mathbb{N} \mid 4 < a < 12\}$?

Solution: Let us let $A = \{a \in \mathbb{N} \mid a \text{ is odd}\}$ and $B = \{a \in \mathbb{N} \mid 4 < a < 12\}$, so that A is the set of elements in $\mathbb{N}$ that are odd (namely, the elements 1, 3, 5, 7, 9, 11, and so on) and B is the set of elements in $\mathbb{N}$ that are greater than 4 and less than 12. In particular, we see that we can explicitly write B as $B = \{5, 6, 7, 8, 9, 10, 11\}$. The question wants us to find the intersection $A \cap B$ of the sets A and B, which means that we need to find all of the elements that are in **both** A and B. We can now see that this means we must find all of the elements in $\mathbb{N}$ that are odd (so that they are in A) **as well as** larger than 4 and less than 12 (so that they are in B). To do this, we can simply look at $B = \{5, 6, 7, 8, 9, 10, 11\}$ and pick out all of the odd elements. Thus, we see that $A \cap B = \{5, 7, 9, 11\}$ is the set of elements that are odd as well as greater than 4 and less than 12.

4) Let A be any set at all, and let O be a subset of A. What is $A \cup O$? What is $A \cap O$?

Solution: If O is a subset of A, then $A \cup O$ is equal to the set A, and $A \cap O$ is equal to the set O. In words, we can read the previous sentence as saying that if O is a subset of A, then the union of A and O is just A, and the intersection of A and O is just O. To see why this is the case, we just need to unpack all of the relevant definitions. Namely, if O is a subset of A then all of the elements in O are also in A (by definition of a subset). Therefore, the set of elements that are in either A or O is exactly the same as the set of elements that are in A, simply because there are no elements in O that are not also in A. Thus, since the union of A and O is the set of elements that are in either A or O, we see that the union of A and O is just A. In symbols, we have that $A \cup O = A$.

Similarly, the set of elements that are in both A and O is exactly the same as the set of elements that are in O. This is because one way to construct the intersection of A and O is to go through every element in O and ask which ones are also in A and place those elements in the intersection.[12] However, since **every** element in O is also in A (since O is a subset of A), we will indeed be placing every element of O in the intersection. In symbols, we have that $A \cap O = O$.

5) Let A and B be any sets at all. Prove that A and B are both subsets of $A \cup B$. Also prove that $A \cap B$ is a subset of A as well as a subset of B.

Solution: We begin by recalling that in order to show that A and B are both subsets of $A \cup B$, we simply need to show that every element in A is also in $A \cup B$, and that every element in B is also in $A \cup B$. However, we recall that $A \cup B$ is **by definition** the set that contains all of the elements that are **either** in A or in B (or in both). Thus, any element in A must be in $A \cup B$ simply by definition of the union, and the same holds for B as well — namely, every element of B must be in $A \cup B$ by definition of the union. This shows that A and B are both subsets of $A \cup B$.

Similarly, by recalling that the intersection of A and B is, by definition, the set of elements that are in **both** A and B, we see that every element in $A \cap B$ must also be in A (since every element of $A \cap B$ is in both A and B). However, this is nothing but the statement that $A \cap B$ is a subset of A. Similarly, every element in $A \cap B$ must also be in B (since every element of $A \cap B$ is in both A and B) and therefore $A \cap B$ is a subset of B as well. We note that even if $A \cap B$ is the empty set, as would happen if $A = \{1, 2, 3\}$ and $B = \{a, b, c, d\}$ for example, the statement that $A \cap B$ is a subset of both A and B is still true. Namely, we already know that the empty set is a subset of **every** set, and so if $A \cap B$ is empty it is still a subset of both A and B just the same.

Both of these proofs are hardly proofs at all — the results that we obtained follow almost immediately from the very definition of unions and intersections. Nonetheless, it is important to check these statements explicitly and see precisely **how** they follow immediately

---

[12]This is one way to ensure that we get **all** of the elements that are in **both** A and O. We could just as well go through every element in A and ask which ones are also in O, as this would be another way to ensure that we get all of the elements that are in both sets.

from the various definitions. This sort of proof — where the desired result follows almost immediately by simply and carefully examining the definitions of the objects involved — is often referred to as "unfolding the definitions." It is always important to know how to unfold definitions, and developing this skill is what motivates this problem.

6) Let A be any set at all. What is the union of A with the empty set? What is their intersection?

Solution: For this we simply steal the results from Exercise 4. Namely, since the empty set is a subset of every set, we can apply the logic of Exercise 4 with O being the empty set itself. In particular, the results of Exercise 4 applied **for any** subset O of A, and since the empty set is indeed a subset of A, those same results apply and we can simply read them off. Namely, the union of any set A with the empty set is simply A again (because the empty set most surely does not "bring anything new to the table"). Similarly, the intersection of any set A with the empty set is the empty set, because there is nothing that is in both A and the empty set, simply because there is nothing in the empty set to begin with!

Solutions To Chapter 17

1) Prove Theorem 17.3.

Solution: Let us first recall that Theorem 17.3 says that for any three sets A, B, and C, we have that

$$(A \cap B) \cup C = (A \cup C) \cap (B \cup C).$$

In order to prove this statement, we proceed as we did for the other identities in this chapter. Namely, we will show that the set on the left is a subset of the set on the right, and then we will show that the set on the right is a subset of the set on the left. In other words, we will show that every element in $(A \cap B) \cup C$ is also in $(A \cup C) \cap (B \cup C)$, and then we will additionally show that every element in $(A \cup C) \cap (B \cup C)$ is also in $(A \cap B) \cup C$. This will establish that these two sets are identical to each other. Let us begin by first showing that every element in $(A \cap B) \cup C$ is also in $(A \cup C) \cap (B \cup C)$.

Let us suppose that x is some arbitrary element in $(A \cap B) \cup C$. This means that x is either in $A \cap B$ or that x is in C (or possibly both). Since our goal is to see whether or not x is also an element of $(A \cup C) \cap (B \cup C)$, let us consider both of these possibilities (namely, the case where x is in $A \cap B$ and the case where x is in C) and see whether or not, in each case, x is an element of $(A \cup C) \cap (B \cup C)$. Let us therefore first suppose that x is in $A \cap B$. Then we know that x is in A and that x is in B, by definition of the set $A \cap B$. Therefore, since x is in A, we know that x is in the set $A \cup C$ because the set $A \cup C$ is by definition the set of elements that are in either A or C. Similarly, since x is in B, we know that x is in the set $B \cup C$ because the set $B \cup C$ is by definition the set of elements that are in either B or C. We therefore see that x is in the set $A \cup C$ as well as the set $B \cup C$. This is precisely the statement that x is in $(A \cup C) \cap (B \cup C)$, since this set is, by definition, the set of elements that are in $A \cup C$ and $B \cup C$. We have therefore shown that if x is in $A \cap B$, then x is in $(A \cup C) \cap (B \cup C)$.

We now need to show that if x is in C, then x is in $(A \cup C) \cap (B \cup C)$. Once we have done this, we will have successfully shown that $(A \cap B) \cup C$ is a subset of $(A \cup C) \cap (B \cup C)$. This is because any element in $(A \cap B) \cup C$ must be in either $A \cap B$ or C, and once we have done this part of the proof we will have shown that in **either** case (i.e., if the element is in **either** $A \cap B$ or C) then that element will also be in $(A \cup C) \cap (B \cup C)$. So suppose that x is an element in C. We then know that x is in $A \cup C$ since $A \cup C$ is the set of elements that are in either A or C. By the exact same reasoning, we also know that x is in $B \cup C$. Thus, we see that if x is in C, then it is also in $(A \cup C) \cap (B \cup C)$, since x being in C means that x is in $A \cup C$ **and** $B \cup C$. This is precisely what we wanted to show, and so we have completed the proof that $(A \cap B) \cup C$ is a subset of $(A \cup C) \cap (B \cup C)$.

We now set out to show that $(A \cup C) \cap (B \cup C)$ is also a subset of $(A \cap B) \cup C$. By doing this we will have shown that these two sets equal each other, and we will have therefore proven Theorem 17.3. Let us therefore begin by considering some arbitrary element x in $(A \cup C) \cap (B \cup C)$. Since x is in $(A \cup C) \cap (B \cup C)$ it is, by definition, in the set $A \cup C$ **as well as** the set $B \cup C$. We can now study two possibilities — either x is in C or it is not.[13] If x is in C, then it is immediately true that x is in the set $(A \cap B) \cup C$, since this set is by definition the set of elementts that are **either** in $A \cap B$ or in C. Now let us suppose that x is **not** in C, and let us recall that we have already assumed that x is an element of $(A \cup C) \cap (B \cup C)$ so that x is in $A \cup C$ as well as in $B \cup C$. In order for x to be in $A \cup C$ while **not** being in C, it must be the case that x is in A. Similarly, in order for x to be in $B \cup C$ while **not** being in C, it must be the case that x is in B. Thus, we have found that if x is **not** in C, then it is in A and B, and therefore in $A \cap B$.

We have therefore shown that if x is in $(A \cup C) \cap (B \cup C)$, then it is either in C or it is in $A \cap B$. However, if x is either in C or $A \cap B$, then it is in $(A \cap B) \cup C$ by definition of the union of two sets. Thus, we have shown that if x is in $(A \cup C) \cap (B \cup C)$ then it is also in $(A \cap B) \cup C$, and therefore we have shown that $(A \cup C) \cap (B \cup C)$ is a subset of $(A \cap B) \cup C$. Since we have already shown that $(A \cap B) \cup C$ is a subset of $(A \cup C) \cap (B \cup C)$, this means that we have shown that these two sets are equal, and therefore that

$$(A \cap B) \cup C = (A \cup C) \cap (B \cup C).$$

This completes our proof.

2) Use Theorem 17.2 to prove that

$$(A \cup B) \cap (C \cup D) = (A \cap C) \cup (A \cap D) \cup (B \cap C) \cup (B \cap D).$$

Hint: We will have to use Theorem 17.2 twice, and in the first instance it is important to view one of the two unions on the left side of the equation as a single set when applying the theorem. (Note: This one is hard. Try to gain some intuition for what both sides of the equality mean in words, and that will help guide the symbolic manipulation.)

---

[13] This is always the case. Namely, any element is either in, or not in, any given set.

Solution: As the footnote to this problem in the main text states, it is important for us to remain calm when attacking this problem. Namely, the long mathematical expression with nested parentheses and an explosion of "∪" and "∩" symbols may seem daunting when looking at the whole thing from a distance, but if we recall that we know what all of these symbols mean and how they relate to each other, we see that we can make good sense of all of this. In particular, all that the expression

$$(A \cup B) \cap (C \cup D) = (A \cap C) \cup (A \cap D) \cup (B \cap C) \cup (B \cap D)$$

really means in words is that "the set of elements that are in 'A or B' as well as in 'C or D' is equal to the set of elements that are in 'A and C' or in 'A and D' or in 'B and C' or in 'B and D'." This is a lesson we should take with us throughout our entire mathematical careers — any time that we see an expression that looks daunting, we just need to take a deep breath, stay calm, and work through it slowly but surely. Namely, every symbol and parenthesis in this expression means something, the exact same way that every letter on this page means something. All we need to do is go through it all, reading each phrase letter by letter and word by word.

The following proof will be largely symbolic in nature. This is meant to show how powerful our main theorems are, and how we can use them to prove statements that would be difficult to prove without the framework and notation that we have introduced so far. Namely, the statement that "the set of elements that are in 'A or B' as well as in 'C or D' is equal to the set of elements that are in 'A and C' or in 'A and D' or in 'B and C' or in 'B and D'," which is what we are about to prove, would be difficult to prove without the notation that we have introduced as this notation allows us to easily manipulate the logical content of this statement in ways that words simply cannot. Indeed, one of the main utilities of mathematics is that it allows us to isolate and manipulate logical statements in very succinct ways, so that we can deduce the validity of very large logical conclusions by first breaking them up into smaller, more digestible truths. Let us now stop the philosophizing and see this all in action by turning our attention towards proving the above statement.

We begin by listening to the hint that was given to us. If we look

SOLUTIONS TO CHAPTER 17

at the set $(A \cup B) \cap (C \cup D)$ we see that there are two unions and one intersection that we need to deal with, and this seems somewhat painful. However, let us for the moment denote the set $C \cup D$ by a single symbol (since $C \cup D$ is, after all, a single set). For concreteness, let us use F to denote $C \cup D$, so that $F = C \cup D$. We then have, by definition of F, that

$$(A \cup B) \cap (C \cup D) = (A \cup B) \cap F.$$

This is a nicer expression for us to deal with because now there is only one union and one intersection that we need to consider, and indeed Theorem 17.2 gives us the tool that we need for dealing with such an expression.

Namely, Theorem 17.2 tells us that for **any** three sets A, B, and C,

$$(A \cup B) \cap C = (A \cap C) \cup (B \cap C).$$

We can therefore immediately apply this theorem to our current case, where instead of having the three sets A, B, and C, we have the three sets A, B, and F. We therefore have that

$$(A \cup B) \cap F = (A \cap F) \cup (B \cap F),$$

where all we did was replace C with F in the expression given to us by Theorem 17.2. If we now plug $C \cup D$ back in for F in the above equation (since these two sets are, by definition of F, equal to each other), we immediately have that

$$(A \cup B) \cap (C \cup D) = (A \cap (C \cup D)) \cup (B \cap (C \cup D)),$$

where all we did was replace F with $C \cup D$ everywhere in the equation given above.

Let us now look at the two expressions $A \cap (C \cup D)$ and $B \cap (C \cup D)$ which both appear on the right hand side of the above equation. Both of these are of the same form — namely, one set intersected with the union of two other sets. We can therefore apply Theorem 17.2 to both of these. Namely, we can apply this theorem[14] to $A \cap (C \cup D)$ to find that

$$A \cap (C \cup D) = (A \cap C) \cup (A \cap D),$$

---

[14] As well as the previous unmentioned (but hopefully clear) facts that, for any

and similarly we can apply this theorem to $B \cap (C \cup D)$ to find that

$$B \cap (C \cup D) = (B \cap C) \cup (B \cap D).$$

All we need to do now is plug these two results into the equation that we proved above, namely

$$(A \cup B) \cap (C \cup D) = (A \cap (C \cup D)) \cup (B \cap (C \cup D)).$$

By plugging
$$A \cap (C \cup D) = (A \cap C) \cup (A \cap D)$$
and
$$B \cap (C \cup D) = (B \cap C) \cup (B \cap D)$$

into the right hand side of the above equation, we find that

$$(A \cup B) \cap (C \cup D) = (A \cap C) \cup (A \cap D) \cup (B \cap C) \cup (B \cap D),$$

as desired. This completes our proof.

As a quick aside, let us note a fascinating connection between the result we just established and a common result from genuine arithmetic. Namely, we have just proved that for any sets A, B, C, and D, we have that

$$(A \cup B) \cap (C \cup D) = (A \cap C) \cup (A \cap D) \cup (B \cap C) \cup (B \cap D).$$

Let us also recall[15] that for any four **numbers** a, b, c, and d, we have that

$$(a + b) \times (c + d) = (a \times c) + (a \times d) + (b \times c) + (b \times d).$$

---

sets A and B, we have that $A \cup B = B \cup A$ and that $A \cap B = B \cap A$. These facts can be quickly confirmed upon reflection on the definition of unions and intersections. Namely, the fact that we write the union or intersection of A and B as $A \cup B$ or $A \cap B$ instead of $B \cup A$ or $B \cap A$ is purely an arbitrary convention, since constructing the union or intersection of A and B keeps both sets on completely equal footing.

[15]If the following algebraic result is not too familiar then that is okay — the discussion of this paragraph is not at all essential for the ideas that we will continue to develop.

SOLUTIONS TO CHAPTER 17   273

For example, if our four numbers are 2, 3, 6, and 9, then we have that $(2+3) \times (6+9) = 75$ (which corresponds to the left hand side of the above equation), and we also have that

$$(2 \times 6) + (2 \times 9) + (3 \times 6) + (3 \times 9) = 75,$$

(which corresponds to the right hand side of the above equation). Now, if we replace the symbol "+" with the symbol "∪" and the symbol "×" with the symbol "∩" (as well as lower case letters representing numbers with capital letters representing sets), then the expression

$$(a+b) \times (c+d) = (a \times c) + (a \times d) + (b \times c) + (b \times d)$$

is identical to the expression

$$(A \cup B) \cap (C \cup D) = (A \cap C) \cup (A \cap D) \cup (B \cap C) \cup (B \cap D).$$

It is important to note, however, that this is a **purely symbolic** relationship. Namely, the first of these expressions is a relationship amongst **numbers** (and how they are added and multiplied), whereas the second of these expressions is a relationship amongst **sets** (and how they are unioned and intersected). We will see in a later volume what the significance of this similarity really is, but for now let us simply note the similarity and how miraculous it is that two seemingly very different processes (namely, that of adding and multiplying numbers and that of unioning and intersecting sets) can indeed be so closely related.

3) **Use Theorem 17.3 to prove that**

$$(A \cap B) \cup (C \cap D) = (A \cup C) \cap (A \cup D) \cap (B \cup C) \cap (B \cup D).$$

The same hint and note from the previous problem apply to this problem as well.

Solution: This problem is almost identical to the previous one, so let us attack it without much of an introduction. However, let us at least explain what this huge chain of symbols actually means in words, for this will help us see that all of this is actually not too scary. Namely, the equation

$$(A \cap B) \cup (C \cap D) = (A \cup C) \cap (A \cup D) \cap (B \cup C) \cap (B \cup D)$$

is simply telling us that "the set of elements that are either in 'A and B' or in 'C and D' is the same as the set of elements that are in 'A or C' as well as in 'A or D' as well as in 'B or C' as well as in 'B or D'." Let us now go on to actually prove this statement.

First, let us begin by reminding ourselves that Theorem 17.3 tells us that for any three sets A, B, and C, we have that

$$(A \cap B) \cup C = (A \cup C) \cap (B \cup C).$$

Since we were told that the hint from the previous problem still applies to this problem, we are motivated to let F be the set $C \cap D$ so that we can write

$$(A \cap B) \cup (C \cap D) = (A \cap B) \cup F.$$

We can now apply Theorem 17.3 directly to this equation, where we simply replace C in the statement of the theorem with F. Thus, the theorem tells us that

$$(A \cap B) \cup F = (A \cup F) \cap (B \cup F),$$

and by plugging this result into the equation

$$(A \cap B) \cup (C \cap D) = (A \cap B) \cup F$$

that we just found, we see that we have

$$(A \cap B) \cup (C \cap D) = (A \cup F) \cap (B \cup F).$$

If we now look at $A \cup F$ and recall that we defined F so that $F = C \cap D$, then we see that we have

$$A \cup F = A \cup (C \cap D).$$

We can therefore apply Theorem 17.3 to the right hand side of this equation again to find that

$$A \cup F = (A \cup C) \cap (A \cup D).$$

In the exact same way, we can replace F with $C \cap D$ in the set $B \cup F$ to find that
$$B \cup F = B \cup (C \cap D),$$
and then we can apply Theorem 17.3 to the right hand side of this equation again to find that
$$B \cup F = (B \cup C) \cap (B \cup D).$$
If we now plug the equations (that we just established)
$$A \cup F = (A \cup C) \cap (A \cup D)$$
as well as
$$B \cup F = (B \cup C) \cap (B \cup D)$$
into the equation
$$(A \cap B) \cup (C \cap D) = (A \cup F) \cap (B \cup F)$$
that we found above, then we see that we have found that
$$(A \cap B) \cup (C \cap D) = (A \cup C) \cap (A \cup D) \cap (B \cup C) \cap (B \cup D),$$
as desired. This completes our proof.

Solutions To Chapter 18

1) Let $A = \{1, 2\}$, $B = \{a, b\}$, and $C = \{3, 4\}$. Then the Cartesian product $A \times B$ of A and B is a perfectly good set, and so we can take its Cartesian product with C, denoted by $(A \times B) \times C$. Write down one example of an element in the set $(A \times B) \times C$. How many elements are in $(A \times B) \times C$?

Solution: Let us begin by exploring how many elements there are in $(A \times B) \times C$, and then we will turn to the issue of writing down the elements in this set explicitly. We claim that there are eight elements in $(A \times B) \times C$. To see this, we must recall that the Cartesian product of a set with N elements and a set with M elements is a set with NM elements — this is the result obtained in Proposition 18.2.

Accordingly, we see that the set $A \times B$ has four elements, since it is the Cartesian product of a 2-element set with a 2-element set, and $2 \times 2 = 4$. The set $(A \times B) \times C$ is then the Cartesian product of $A \times B$ with C, which is the Cartesian product of a 4-element set with a 2-element set. Thus, the set $(A \times B) \times C$ has eight elements, which is precisely what we claimed above.

To be even more explicit, and to check that we are right, we can simply write out all of the elements in $(A \times B) \times C$. In doing so, we will also answer the part of this exercise which asks us to write down one example of an element in this set, for we will indeed write down **every** element in this set. To do this, we first need to know all of the elements in $A \times B$. These are simply $(1, a)$, $(1, b)$, $(2, a)$, and $(2, b)$. It is now these four elements that make an appearance in the "first slot" of the pairs in the set $(A \times B) \times C$. Namely, just as the elements in A are placed in the "first slot" of the pairs in $A \times B$ (with the elements of B being placed in the "second slot"), we need to place the elements of $A \times B$ in the "first slot" of the pairs in $(A \times B) \times C$, and the elements of C in the "second slot." Now, however, the elements that go in the "first slot" are themselves pairs, but this is not a problem — all we have to do is employ nested parentheses just as we are already accustomed to. Namely, to find the elements of $(A \times B) \times C$, we simply take each element of $A \times B$ and place it in the "first slot," and then place either 3 or 4 in the "second slot."

For example, if we take the element $(1, a)$ from $A \times B$ and the element 3 from C, we can form the element $((1, a), 3)$ in the set $(A \times B) \times C$. We note that this is in line with everything we have said so far about Cartesian products — namely, an element of the left hand set in the Cartesian product is in the "first slot" of our pair, and an element of the right hand set is in the "second slot." Doing this for all possible combinations of elements in $A \times B$ and C, we see that the eight elements of $(A \times B) \times C$ are $((1, a), 3)$, $((1, a), 4)$, $((1, b), 3)$, $((1, b), 4)$, $((2, a), 3)$, $((2, a), 4)$, $((2, b), 3)$, and $((2, b), 4)$. Indeed, there are eight elements in total.

2) Let $A = \{1, 2, 3\}$. Write down all of the elements in $A \times A$. How many elements are in $A \times A$?

SOLUTIONS TO CHAPTER 18

Solution: This exercise is actually a bit more straightforward than the previous one — the main point of it is to point out that we can indeed take Cartesian products of sets with themselves (just as we can consider the union and/or intersection of a set with itself). Let us therefore dive right in. We will first answer the question of how many elements there are in $A \times A$, and then we will go on to explicitly write out all of the elements in this set.

Since A is a 3-element set, the Cartesian product of A with itself is a 9-element set. This is again because the Cartesian product of a set with N elements and a set with M elements is a set that has NM elements. Here, both N and M are three, since the two sets that we are taking the Cartesian product of are both A, which has three elements. This answers the second part of this problem.

To check this, we can list these nine elements explicitly. By creating **all possible** pairs (a, b) where **both** a and b are elements in A, we find that the elements in $A \times A$ are

$$(1,1), (1,2), (1,3), (2,1), (2,2), (2,3), (3,1), (3,2), \text{ and } (3,3).$$

One way of seeing this is to first choose 1 in A as the element in the "first slot," and then "cycle through" all three of the elements in A for the "second slot." We then do the same with the element 2 in A being in the "first slot," and then again with the element 3 in A being in the "first slot." This ensures that we do not miss any elements, and this also completes the first part of this exercise.

Now that we have completed both parts of the exercise, let us note a couple of interesting points. We mentioned above that we can also take unions as well as intersections of sets with themselves. Namely, we can consider $A \cup A$ as well as $A \cap A$, just as we have considered $A \times A$ in this exercise. However, it is important to note that the results of doing these three things — namely, of taking the union of a set with itself, of taking the intersection of a set with itself, and of taking the Cartesian product of a set with itself — are vastly different.

The union and intersection of any set with itself are both simply the original set. We can see this by noting that $A \cup A$ is the set of all elements that are in either A or A, but this set is simply A again. Thus, we have $A \cup A = A$. Similarly, the set $A \cap A$ is the set of all elements that are in A **and** A, which we again see is simply A again. Thus, we have $A \cap A = A$.

However, the Cartesian product of a set with itself is vastly different from the original set, as we have seen in this exercise. Namely, $A \times A \neq A$. The reason for this comes down to the fact that, as mentioned in the main text of this chapter, in order to construct the Cartesian product of two sets we must **change the elements of the sets**, whereas in the union and intersection constructions the elements themselves are left unchanged. In particular, the elements in the Cartesian product are **pairs** of elements in the original sets, and not the elements themselves. Let us note also that $(1, 2)$ and $(2, 1)$ are **different elements** in the set $A \times A$. One way to make sense of this (if it does not already make sense) is to think of the elements (a, b) as "ways of choosing a from the first set and b from the second set." We then see that picking 1 from the first set and 2 from the second set is indeed a different choice than picking 2 from the first set and 1 from the second set. Thus, these two different "ways of choosing elements" should correspond to two different elements in $A \times A$. The fact that $(1, 2) \neq (2, 1)$, i.e. that $(1, 2)$ and $(2, 1)$ are different elements in $A \times A$, does indeed reflect this.

3) **Let $A = \{1, 2\}$, $B = \{a, b\}$, and $C = \{b, c, d\}$. Write down all of the elements in the set $A \times (B \cup C)$. How many should there be? Also write down all of the elements in the set $A \times (B \cap C)$. How many should there be in this case?**

Solution: As we have done before, let us begin by asking about the number of elements in the sets $A \times (B \cup C)$ and $A \times (B \cap C)$ before we go on to explicitly write down these elements. We note that since the number of elements in any Cartesian product $A \times B$ is the number of elements in A times the number of elements in B (assuming both A and B have a finite number of elements), we have that the number of elements in $A \times (B \cup C)$ is equal to the number of elements in A times the number of elements in $B \cup C$. These two sets are, after all, the sets that we are taking the Cartesian product of, and therefore the number of elements in each of these two sets will determine the number of elements in the set $A \times (B \cup C)$. Similarly,

SOLUTIONS TO CHAPTER 18

the number of elements in the set $A \times (B \cap C)$ will be the number of elements in A times the number of elements in $B \cap C$.

We therefore need to figure out how many elements there are in $B \cup C$, as well as the number of elements in $B \cap C$. Once we have these pieces of information, we can use the fact that A has two elements to finish this part of the problem. Fortunately, constructing unions and intersections of two sets is something we have already done plenty of. Namely, we see that the union of B and C is simply $\{a, b, c, d\}$, where we note that we do not "double count" the element b even though it is in both B and C. Thus, $B \cup C = \{a, b, c, d\}$ and we therefore see that $B \cup C$ has four elements. Since A has two elements, we have that $A \times (B \cup C)$ has $2 \times 4 = 8$ elements.

Similarly, since the only element that is in both B and C is b, we see that $B \cap C = \{b\}$ and therefore we see that $B \cap C$ only has one element. Thus, we have that $A \times (B \cap C)$ has $2 \times 1 = 2$ elements.

We can check all of this explicitly by writing out the elements in $A \times (B \cup C)$ as well as the elements in $A \times (B \cap C)$. Let us begin with $A \times (B \cup C)$. We have that $A = \{1, 2\}$ and that $B \cup C = \{a, b, c, d\}$. We therefore get four elements by choosing 1 in A as the element in the "first slot" and "cycling through" the four elements in $B \cup C$. Namely, we get the four elements $(1, a)$, $(1, b)$, $(1, c)$, and $(1, d)$. We now need to move on to the case when 2 in A is in the "first slot," and again "cycle through" every element in $B \cup C$. We then get the four elements $(2, a)$, $(2, b)$, $(2, c)$, and $(2, d)$. These are all of the possible elements in $A \times (B \cup C)$ since we have put each element of A in the "first slot" and in each case "cycled through" all of the elements in $B \cup C$. Thus, since each of the elements in the two groups of four that we just wrote down are distinct from each other, we therefore see that $A \times (B \cup C)$ has eight elements in total, just as expected.

Let us complete this problem by explicitly writing down all of the elements in the set $A \times (B \cap C)$. We have that $A = \{1, 2\}$ and that $B \cap C = \{b\}$. Thus, the only possible entry in the "second slot" is b, and so we can immediately see that the elements in $A \times (B \cap C)$ are simply $(1, b)$ and $(2, b)$. This completes the problem.

One interesting thing to note here is that, by simply inspecting the elements of $A \times (B \cup C)$ and $A \times (B \cap C)$, we see that $A \times (B \cap C)$ is a subset of $A \times (B \cup C)$. Namely, every element in $A \times (B \cap C)$ is also an element of $A \times (B \cup C)$, which is easy to see because $(1,b)$ and $(2,b)$ are the only two elements in $A \times (B \cap C)$ and both of these elements are in $A \times (B \cup C)$. Let us leave it as an extra to challenge for us to consider why we could have known that this would be the case before we even explicitly wrote down the various elements. Namely, how can we see that $A \times (B \cap C)$ is a subset of $A \times (B \cup C)$ without ever writing down a single element?

Solutions To Chapter 19

1) Let $A = \{1,2\}$, $B = \{2,3\}$, and $C = \{1,3\}$. **How many elements are in $(A \sqcup B) \sqcup C$? How many elements are in $A \sqcup (B \sqcup C)$? Are the sets $(A \sqcup B) \sqcup C$ and $A \sqcup (B \sqcup C)$ equal? (This is a very subtle question, and is indeed a warm-up for the mysterious category theory that we have been alluding to for a while. Therefore solving this problem is not so important at this stage, but understanding its solution is quite enlightening.).**

Solution: The solution to this problem is extremely important. In particular, after we discuss the number of elements in the various particular sets that this exercise asks about, we will go on to discuss the subtle issue of taking the disjoint union of three sets. Closely related to this issue is the issue of taking the Cartesian product of three sets, and so we will indeed look at the analogous issues for this construction as well. These concepts will be important for us in the remaining exercises for this chapter, as well as in the next volume, as well as much further down the road when we study categories. Accordingly, we will spend a good deal of time discussing these issues here. However, let us first begin with the more straightforward task of counting elements.

We begin by recalling from the main text of this chapter that the disjoint union of a set with N elements and a set with M elements is a set with $N + M$ elements. Thus, the disjoint union of A and B has four elements, simply because both A and B have two elements. Now, if we take the disjoint union of the set $A \sqcup B$ with the set C to form the set $(A \sqcup B) \sqcup C$, then we are taking the disjoint union of a 4-element set with a 2-element set to form a 6-element set. In short, $(A \sqcup B) \sqcup C$ has six elements. The exact same logic applies to the set $A \sqcup (B \sqcup C)$, so we see that this set also has six elements. Namely, since both B and C are 2-element sets, we have that $B \sqcup C$ has $2 + 2 = 4$ elements. We are then taking the disjoint union of a 2-element set (in this case, A) with a 4-element set (in this case, $B \sqcup C$), to obtain a 6-element set.

Subtlety With Disjoint Unions

In order to investigate the question

$$(A \sqcup B) \sqcup C \stackrel{?}{=} A \sqcup (B \sqcup C),$$

we need to get a bit more detailed. Namely, the question of whether or not $(A \sqcup B) \sqcup C$ and $A \sqcup (B \sqcup C)$ are equal requires us to see whether or not the elements of these sets are identical, and this in turn requires us to ask what the elements of these sets even are. In order to do this we are going to very strictly stick to the notation developed in this chapter. Let us be warned, however, that this will get a little bit ugly. That said, let us recall that it is just notation, and we should never let notation scare us! All we need to do is take a deep breath and remember what the notation means.

Let us begin by trying to understand the elements of $(A \sqcup B) \sqcup C$. To do this, we first need to write out the elements in $A \sqcup B$. Using what we learned in this chapter and noting that $A = \{1, 2\}$ and $B = \{2, 3\}$ as given in this exercise, we see that we have

$$A \sqcup B = \{(1, A), (2, A), (2, B), (3, B)\}.$$

This is simply because we put the element in the "first slot" and the set that "labels" the element in the second slot. Note, for example, that $(2, A)$ and $(2, B)$ are **different elements** in the set $A \sqcup B$ even though the element 2 that shows up in A is identical to the element 2 that shows up in B. This is because neither $(2, A)$ nor $(2, B)$ are equal to the element 2. Rather, the elements $(2, A)$ and $(2, B)$ are each the element 2 **with an extra label** (namely, a label of the set from which it comes). This is a more abstract element than simply the number 2.

Now, in order to write out the elements in $(A \sqcup B) \sqcup C$, we simply do this again. Namely, there are two types of elements in $(A \sqcup B) \sqcup C$. There are those that come from the set $A \sqcup B$ and those that come from the set C. Just as we do for any disjoint union, we must write the elements in $(A \sqcup B) \sqcup C$ that come from $A \sqcup B$ as a pair, where the **element** in $A \sqcup B$ is placed in the "first slot" and the **set** $A \sqcup B$ itself is placed in the "second slot," so as to "label" the elements in the set. Similarly, we must write the elements in $(A \sqcup B) \sqcup C$ that come from C as a pair, where the element in C is placed in the "first slot" and the set C itself is placed in the "second slot," so as to "label" the elements in the set.[16]

Let us put this all together and simply write out all of the elements in $(A \sqcup B) \sqcup C$, and then we will follow this up with even more explanation (thus, if the following equation makes no sense then do not worry, we will try to further explain it after we state it). Namely, we have

$$(A \sqcup B) \sqcup C = \{((1,A), A \sqcup B), ((2,A), A \sqcup B), ((2,B), A \sqcup B),$$
$$((3,B), A \sqcup B), (1,C), (3,C)\}.$$

Let us take a closer look at the element $((1, A), A \sqcup B)$ and try to understand why this element is written this way based solely on the reasoning we gave above. Namely, as we said above, we want the "first slot" of this element to be an element of $A \sqcup B$. Indeed, $(1, A)$ is a **single element** in $A \sqcup B$, and it is taking up the "first slot" of $((1, A), A \sqcup B)$. Additionally, we want to "label" these elements by the set from which they come by placing that set in the "second slot" of the relevant elements. In this case, the element $(1, A)$ comes from the set $A \sqcup B$, and this is why the expression $A \sqcup B$ is placed in the "second slot" of $((1, A), A \sqcup B)$.

We also obtain the three elements $((2, A), A \sqcup B)$, $((2, B), A \sqcup B)$, and $((3, B), A \sqcup B)$ by applying the same reasoning to the elements $(2, A)$, $(2, B)$, and $(3, B)$ in $A \sqcup B$. Namely, we place the **elements** in the first slot and the **name of the set** in the second slot.

---

[16] We note that this is precisely what we do in the case of a single disjoint union. For example, the elements in the set $A \sqcup B$ are precisely of this form, only in our current case the set A is being replaced by the set $A \sqcup B$ and the set B is being replaced by the set C.

Let us now take a closer look at the elements $(1, C)$ and $(3, C)$ in the set $(A \sqcup B) \sqcup C$. We see that these elements come from applying the same logic that we used for the set $A \sqcup B$ to the set C. Namely, 1 and 3 are elements in C and so we get two new elements — one with 1 in the "first slot" and C in the "second slot," and one with 3 in the "first slot" and C in the "second slot."

Thus, we see that the only difference between the elements in $(A \sqcup B) \sqcup C$ that come from $A \sqcup B$ and those that come from C is that the elements of $A \sqcup B$ are already pairs — they are elements in A or in B and are labeled by the set from which they come. Namely, the elements in $A \sqcup B$ are pairs like $(1, A)$ or $(2, B)$. However, the elements in C are simply numbers. In either case, though, in order to create the elements that make up $(A \sqcup B) \sqcup C$, we need to take each of the elements in $A \sqcup B$ and label them by $A \sqcup B$, and then take each of the elements in C and label them by C. This is how we obtained the elements that we listed above, and we note that it follows from nothing but a precise and careful following of the definition of the disjoint union.

Let us recall that we are interested in writing out these elements explicitly because we want to answer the question

$$(A \sqcup B) \sqcup C \stackrel{?}{=} A \sqcup (B \sqcup C).$$

Thus, we now need to turn our attention to the set $A \sqcup (B \sqcup C)$ and try to explicitly write down its elements. This will allow us to compare the elements from these two sets and determine if they are the same. Fortunately, after being so careful with the set $(A \sqcup B) \sqcup C$, our work with the set $A \sqcup (B \sqcup C)$ will be easier.

As we did before, we need to begin by examining the set $B \sqcup C$. Namely, by recalling that $B = \{2, 3\}$ and $C = \{1, 3\}$, we see that

$$B \sqcup C = \{(2, B), (3, B), (1, C), (3, C)\}.$$

Now, in order to write down the elements in $A \sqcup (B \sqcup C)$, we need to first write down all of the elements in A with A as a label. For example, $(1, A)$ is an element of $A \sqcup (B \sqcup C)$. We then need to write down all of the elements of $B \sqcup C$ with $B \sqcup C$ as a label. For example, $((1, C), B \sqcup C)$ is an element of $A \sqcup (B \sqcup C)$. Putting these all together, we have that

$$A \sqcup (B \sqcup C) = \{(1, A), (2, A), ((2, B), B \sqcup C),$$
$$((3, B), B \sqcup C), ((1, C), B \sqcup C), ((3, C), B \sqcup C)\}.$$

We are now in a position to answer the question

$$(A \sqcup B) \sqcup C \stackrel{?}{=} A \sqcup (B \sqcup C).$$

Indeed, as things stand, the answer to this question is no — these two sets are **not** equal. The reason for this is clear, for we simply notice that the elements in these two sets are different. For example, the element $((3, B), B \sqcup C)$ is in the set $A \sqcup (B \sqcup C)$ but is not in the set $(A \sqcup B) \sqcup C$, and similarly the element $((2, A), A \sqcup B)$ is in the set $(A \sqcup B) \sqcup C$ but is not in the set $A \sqcup (B \sqcup C)$.

This is a highly unsatisfactory state of affairs, for a couple of reasons. First, we have already seen (in Chapter 17) that in the case of unions and intersections, we have

$$(A \cup B) \cup C = A \cup (B \cup C)$$

as well as
$$(A \cap B) \cap C = A \cap (B \cap C),$$

while in the case of disjoint unions we have just found that

$$(A \sqcup B) \sqcup C \neq A \sqcup (B \sqcup C).$$

Although there is nothing **inherently wrong** about this — namely, these equalities (and non-equalities) follow from the definition of unions, intersections, and disjoint unions, and so we have simply shown that these constructions behave differently once three sets are involved — it is indeed slightly unappealing that the disjoint union does not share such a basic property with unions and intersections. Namely, it is unappealing that the **order** in which we construct disjoint unions matters, whereas for unions and intersections the order does not.

A **much more potent** reason for why the fact that $(A \sqcup B) \sqcup C$ and $A \sqcup (B \sqcup C)$ are unequal is a problem is the following. We see that, as things stand, the element $((2, B), A \sqcup B)$ in $(A \sqcup B) \sqcup C$ is not equal to the element $((2, B), B \sqcup C)$ in $A \sqcup (B \sqcup C)$. These elements are not the same simply because of how they are written — one element is labeled by $A \sqcup B$ and the other element is labeled by $B \sqcup C$. However, the "actual information" of the elements $((2, B), A \sqcup B)$ as well as $((2, B), B \sqcup C)$ is simply the element 2, coming from B. Namely, it is the "first slot" alone that tells us the useful information here — the labels $A \sqcup B$ and $B \sqcup C$ are not distinguishing the element $(2, B)$ from any other appearance of $(2, B)$ in their respective sets. For example, "$(2, B)$" does not show up anywhere in the set $A \sqcup (B \sqcup C)$ **besides** in the element $((2, B), B \sqcup C)$. Thus, the label $B \sqcup C$ is not really doing anything for us in this case.

To make all of this more clear, we need to go back and consider what it is we really wanted from our definition of a disjoint union. Namely, we sought to define the construction of the disjoint union so that we could "remember" which set a given element comes from. Namely, if an element a is in both A and B, then a had **two separate** appearances in the set $A \sqcup B$ in the forms of (a, A) and (a, B). In this way we do not "lose out" on any information when we take the disjoint union of A and B, the way that we would if we considered $A \cup B$ where the element a only makes one appearance.

We can now notice that the expressions $(1, A)$, $(2, A)$, $(2, B)$, $(3, B)$, $(1, C)$, and $(3, C)$ all make precisely one appearance in both $(A \sqcup B) \sqcup C$ as well as $A \sqcup (B \sqcup C)$. For example, $(1, A)$ makes its appearance in $(A \sqcup B) \sqcup C$ in the form of $((1, A), A \sqcup B)$, whereas it makes its appearance in $A \sqcup (B \sqcup C)$ in the form of $(1, A)$ itself. Similarly, $(2, B)$ makes its appearance in $(A \sqcup B) \sqcup C$ in the form of $((2, B), A \sqcup B)$, whereas it makes its appearance in $A \sqcup (B \sqcup C)$ in the form of $((2, B), B \sqcup C)$.

It is indeed promising that these six elements make exactly one appearance (albeit in different forms) in both $(A \sqcup B) \sqcup C$ as well as $A \sqcup (B \sqcup C)$, for the following reason. Just as we want $A \sqcup B$ to "remember" which elements come from A and which come from B, we similarly would want a "three-fold disjoint union" of A, B, and C to "remember" which elements come from A, which come from B, and which come from C. In some sense, the sets $(A \sqcup B) \sqcup C$ and $A \sqcup (B \sqcup C)$ satisfy this "remembering" property, because each of the six elements $(1, A)$, $(2, A)$, $(2, B)$, $(3, B)$, $(1, C)$, and $(3, C)$ appear only once in both sets.

As things stand, however, the two different appearances of these elements are not the same — namely, $(1, A)$ is not equal to $((1, A), A \sqcup B)$ and $((2, B), A \sqcup B)$ is not equal to $((2, B), B \sqcup C)$. This is because we can think of the element $((1, A), A \sqcup B)$ as telling us that $(1, A)$ made its way into $(A \sqcup B) \sqcup C$ by first "going through" the set $A \sqcup B$, whereas the element $(1, A)$ made its way into $A \sqcup (B \sqcup C)$ directly from A. Similarly, we can think of the element $((2, B), A \sqcup B)$ as telling us that $(2, B)$ made its way into $(A \sqcup B) \sqcup C$ by first "going through" the set $A \sqcup B$, whereas the element $((2, B), B \sqcup C)$ made its way into $A \sqcup (B \sqcup C)$ by first "going through" the set $B \sqcup C$.

The key point to this whole discussion is that if we want to consider a "three-fold disjoint union" of the sets A, B, and C, where the final set remembers which elements come from A, which come from B, and which come from C, it **should not matter** whether some given element first "goes through" the set $A \sqcup B$ or if it first "goes through" the set $B \sqcup C$, for example. This is because we only care about remembering whether or not some given element in our three-fold disjoint union comes from A, B, or C.

Therefore, the sets $(A \sqcup B) \sqcup C$ and $A \sqcup (B \sqcup C)$ are in a sense giving us **too much** information. Namely, both of these sets are indeed "remembering" which elements come from A, which come from B, and which come from C, and this is reflected by the fact that in both sets the six expressions $(1, A)$, $(2, A)$, $(2, B)$, $(3, B)$, $(1, C)$, and $(3, C)$ all make exactly one appearance. However, these two sets are also keeping the information of whether certain elements first "go through" some intermediate set, like $A \sqcup B$ or $B \sqcup C$. **It is this extra information that is making these two sets not be equal**. For example, it is the **extra information** that $(2, B)$ first "goes through" $A \sqcup B$ (as opposed to first "going through" $B \sqcup C$) which makes the element $((2, B), A \sqcup B)$ not equal to $((2, B), B \sqcup C)$. It is also precisely this information that we just described as not mattering.

Let us therefore define the following notion of a three-fold disjoint union. Namely, suppose A, B, and C are the three sets that are given in this exercise. Let us simply **define** the set $A \sqcup B \sqcup C$ (note the lack of parentheses) to be the set whose elements are the six elements

$(1, A)$, $(2, A)$, $(2, B)$, $(3, B)$, $(1, C)$, and $(3, C)$. In other words, we have

$$A \sqcup B \sqcup C \equiv \{(1, A), (2, A), (2, B), (3, B), (1, C), (3, C)\}.$$

There are a couple of **very** important things that we should note about this whole set up.

Firstly, let us note that the set $A \sqcup B \sqcup C$ as we just defined it does not have any intermediate sets at all — we go directly from the elements of A, B, and C to the elements of $A \sqcup B \sqcup C$. One consequence of this is that our notation has simplified greatly, and we do not need to worry about nested parentheses and unnecessary labels like we did in the cases of $(A \sqcup B) \sqcup C$ and $A \sqcup (B \sqcup C)$.

Secondly, and most importantly, we see that **none of the sets** $A \sqcup B \sqcup C$, $(A \sqcup B) \sqcup C$, **and** $A \sqcup (B \sqcup C)$ **are equal to each other**. However (and this is a very important "however"), **we can define a bijective function from any one of these sets to any other of these sets**. Namely, the obvious bijective function that we can define from any of these sets to any other of these sets is the one that sends the element that corresponds to $(1, A)$ in one set to the element that corresponds to $(1, A)$ in the other set. For example, we can define a bijective function from $A \sqcup B \sqcup C$ to $A \sqcup (B \sqcup C)$ by sending $(1, A)$ in $A \sqcup B \sqcup C$ to $(1, A)$ in $A \sqcup (B \sqcup C)$, $(2, A)$ in $A \sqcup B \sqcup C$ to $(2, A)$ in $A \sqcup (B \sqcup C)$, $(2, B)$ in $A \sqcup B \sqcup C$ to $((2, B), B \sqcup C)$, and so on.

What makes this second point[17] so important is that, in a very precise sense, sets that have bijective functions between them are somewhat "equivalent." Namely, when a bijective function exists between two sets,[18] we can "identify" the elements of the two sets with each other and then almost anything that we can do with one set we can do with the other. For example, the two sets $\{a, b, c\}$ and $\{1, 2, 3\}$ clearly have a bijective function between them (indeed, there

---

[17] Namely, our ability to define bijective functions between any two of the three sets $A \sqcup B \sqcup C$, $(A \sqcup B) \sqcup C$, and $A \sqcup (B \sqcup C)$.

[18] Recall from Proposition 15.1 (in Chapter 15) that we can talk about bijective functions "between" sets (as opposed to having to specify which set the function goes **from** and **to**) because we know that if there is a bijective function one way, then there is also a bijective function the other way.

are many bijective functions between these two sets), and in a very clear way we can view these sets as somehow "equivalent." We leave the word "equivalent" in scare quotes because the sets $\{a, b, c\}$ and $\{1, 2, 3\}$ are certainly not **equal**, but in some sense they are "virtually the same."

The previous paragraph may seem somewhat vague, and unfortunately this will have to remain the case until we study category theory much further down the road. Namely, category theory allows us to **precisely** describe the sense in which sets that have bijective functions between them are "equivalent," and this will then help us **precisely** relate the sets $A \sqcup B \sqcup C$, $(A \sqcup B) \sqcup C$, and $A \sqcup (B \sqcup C)$. For now, let us recap what we have seen.

What we have seen is that **none** of the sets $A \sqcup B \sqcup C$, $(A \sqcup B) \sqcup C$, and $A \sqcup (B \sqcup C)$ are equal to each other. However, we have also seen that they are all in some sense "equivalent." Namely, they all carry the same relevant information, which is that they all have one appearance of each of the six elements $(1, A)$, $(2, A)$, $(2, B)$, $(3, B)$, $(1, C)$, and $(3, C)$. What makes these sets unequal is that some of these elements appear in these sets with extraneous information — information that we do not really care about when constructing a disjoint union of three sets.

With all of this in mind, let us now turn our attention to defining the disjoint union of three sets for any arbitrary A, B, and C, instead of just for the particular A, B, and C that are given in this exercise. In particular, in the case of unions and intersections, we did not have to make the three-fold union or three-fold intersection a new definition. This is because we saw that we could take the sets A, B, and C, and take their union (or intersection) two at a time in any order we like and we will always land on the same set. As we have seen here, however, the same is not true for disjoint unions — choosing different orderings of taking disjoint unions results in different (though "equivalent"' in the vague sense described above) sets. We are, however, free to **define** the three-fold disjoint union in almost any way we like, so long as it properly reflects the properties of disjoint unions that we want in the first place. Namely, let us choose the simplest way to define the three-fold disjoint union so that the resulting set still "remembers" where each element comes from.

We will use the generalization of the set $A \sqcup B \sqcup C$ that we defined above as our definition of a three-fold disjoint union in general. Namely, we suppose that A, B, and C are now **any** sets at all. For every element a in A, let us form the pair (a, A). For example, if $A = \{\text{donkey}, \text{cow}, \text{horse}\}$, then we would form the three pairs (donkey, A), (cow, A), (horse, A). Similarly, let us form the pairs (b, B) for all the elements b in B, and let us form the pairs (c, C) for all the elements c in C. All we are doing here is "labelling" the elements in each set by the set from which it comes, just like we did when we were taking the disjoint union of only two sets.

Let us now **define** the set $A \sqcup B \sqcup C$ to be the set of all of these pairs — namely, $A \sqcup B \sqcup C$ is the set of all pairs coming from the sets A, B, and C. We can readily see that for the particular sets A, B, and C that are given in this exercise, the set $A \sqcup B \sqcup C$ that we defined above is indeed in line with the definition of the general "three-fold disjoint union" that we just made.

This definition of $A \sqcup B \sqcup C$ for any arbitrary sets A, B, and C indeed gives a set that "remembers" where all of its elements come from. Moreover, since we do not need to make any arbitrary choices of which pair of sets to take the disjoint union of first, we are avoiding all of the complications that we discussed above. Namely, the set $A \sqcup B \sqcup C$ has all the same **relevant** information as $(A \sqcup B) \sqcup C$ and $A \sqcup (B \sqcup C)$, but without all of the unnecessary notation and subtlety.

Finally, we see that we can extend this definition to a four-fold, five-fold, six-fold, and indeed N-fold disjoint union (where N is **any** positive whole number greater than 1). For example, if we want to take the four-fold disjoint union of the sets A, B, C, and D, we simply form all the pairs of the form (a, A) with a being an element of A, (b, B) with b being an element of B, (c, C) with c being an element of C, and (d, D) with d being element of D. We then define $A \sqcup B \sqcup C \sqcup D$ to be the set of all of these pairs.

This ends our discussion of some of the subtleties that arise when dealing with many-fold disjoint unions. As mentioned above, these issues will come up in the somewhat near future, and more importantly they will show up in the much more distant future of category theory. For now, these considerations give us a good glimpse of just how precise we must be as mathematicians, as well as provide a good example of how our definitions sometimes lead us into trouble and some of the various ways that we can get out of that trouble.

Let us now go on to see how these considerations also apply to the construction of the Cartesian product. Fortunately, the following considerations are in many ways more straightforward than those of the disjoint union, so we should be well equipped to deal with the issues to come.

Subtlety With Cartesian Products

Let us dive right in. As a concrete example, let us suppose that A, B, and C are as given in this exercise so that we have $A = \{1, 2\}$, $B = \{2, 3\}$, and $C = \{1, 3\}$. Let us now consider the two sets $(A \times B) \times C$, which is constructed by first taking the Cartesian product of A and B to form the set $A \times B$, and then taking the Cartesian product of $A \times B$ with the set C. We will then consider the set $A \times (B \times C)$, which is constructed by first taking the Cartesian product of B and C to form the set $B \times C$, and then taking the Cartesian product of $B \times C$ with the set A. We will then see, similarly to the case of the three-fold disjoint union, that

$$(A \times B) \times C \neq A \times (B \times C).$$

We begin by writing the elements of $A \times B$. As we have done this sort of thing many times before, we will simply quote the result. Namely, we have

$$A \times B = \{(1, 2), (1, 3), (2, 2), (2, 3)\}.$$

We now need to think carefully about the set $(A \times B) \times C$. Namely, this set has two "slots" (as any Cartesian product does). In the "first slot" we place elements of $A \times B$ and in the "second slot" we place elements of C. The set $(A \times B) \times C$ is then the set of all possible such pairs.

The only subtlety that arises here is that the elements of $A \times B$ are **themselves** pairs, so that what goes into the "first slot" of an element in $(A \times B) \times C$ is a pair (namely, from $A \times B$). For example, the element in $(A \times B) \times C$ corresponding to the pair $(1, 2)$ from $A \times B$ and 1 from C is written as $((1, 2), 1)$. The nested parentheses are important here — they are what make it clear that $(1,2)$ is in the "first slot" and that 1 is in the "second slot." Continuing on with all possible pairs from $A \times B$ and C gives the following explicit representation of the set $(A \times B) \times C$.

$$(A \times B) \times C = \{((1,2),1), ((1,3),1), ((2,2),1), ((2,3),1),$$
$$((1,2),3), ((1,3),3), ((2,2),3), ((2,3),3)\}.$$

We note that there are eight elements in $(A \times B) \times C$, which is expected since $A \times B$ has four elements, C has two elements, and the number of elements in the Cartesian product of two finite sets is the product of the number of elements in each constituent set.

Let us now carry out the same analysis for the set $A \times (B \times C)$. We begin by noting that the elements in $B \times C$ are $(2,1), (2,3), (3,1), (3,3)$. Then, to form an element of $A \times (B \times C)$ we must place an element of A in the "first slot" and an element of $B \times C$ in the "second slot." For example, the pair $(1,(2,1))$ is an element in $A \times (B \times C)$, since 1 is an element of A and is in the "first slot," and $(2,1)$ is an element of $B \times C$ and is in the "second slot." We again note that the nested parentheses are important, but we also note that they are nested **differently** than they are in the set $(A \times B) \times C$, and this corresponds to the **order** in which we have chosen to take our Cartesian products.

In particular, let us compare the element $((1,2),1)$ in the set $(A \times B) \times C$ with the element $(1,(2,1))$ in the set $A \times (B \times C)$. We immediately see that these two elements are not the same, since their parentheses are nested differently. By taking all possible pairs of elements in A in the "first slot" and elements of $(B \times C)$ in the "second slot," we see that we have

$$A \times (B \times C) = \{(1,(2,1)), (1,(2,3)), (1,(3,1)), (1,(3,3)),$$
$$(2,(2,1)), (2,(2,3)), (2,(3,1)), (2,(3,3))\}.$$

Indeed, we immediately see that

$$(A \times B) \times C \neq A \times (B \times C).$$

In fact, these sets are "**very** not equal," in the sense that there is **not a single element** that is in both of these sets — all of the parentheses in the elements of $(A \times B) \times C$ are nested differently than those in the elements of $A \times (B \times C)$.

The fact that these two sets are "**very not equal**" is highly troublesome. The reason for this is that there is also a very clear sense in which these sets are "equivalent." For example, **every** element in $(A \times B) \times C$ makes an appearance in $A \times (B \times C)$ if we simply "shift the parentheses over." For example, $((2,3),1)$ is an element of $(A \times B) \times C$ and $(2,(3,1))$ is an element of $A \times (B \times C)$, and $(2,(3,1))$ is identical to $((2,3),1)$ only with the parentheses around 3 and 1 "shifted over" to being around 2 and 3. A moment's reflection shows us that this difference in how we nest our parentheses follows solely from our different choices of which pair of sets we take the Cartesian product of first. In general, we want as little of math as possible to depend on our arbitrary choices, and so it would be nice if we could define a three-fold Cartesian product that does not depend so heavily on our choices.

As things stand, we would need to define only **one of** the sets $(A \times B) \times C$ or $A \times (B \times C)$ as "the" three-fold Cartesian product, since these sets are not equal to each other. However, there is not at all an obvious choice to make — both of these sets (or even some other ordering of which sets to take the Cartesian product of first) seem like equally good candidates, and so we would have to make a **completely arbitrary** choice for which one should be delegated as "the" three-fold Cartesian product.

Before we give a new and more satisfactory definition of the three-fold Cartesian product, let us first note that it is again the case that there is a very natural bijective function going between the two sets $(A \times B) \times C$ and $A \times (B \times C)$. Namely, we can define the function that sends each element to its "parentheses-shifted" partner in the other set. Such a function would, for example, send $((1,2),3)$ in $(A \times B) \times C$ to the element $(1,(2,3))$ in $A \times (B \times C)$. This function is clearly bijective.

Since there is a bijection between the two sets $(A \times B) \times C$ and $A \times (B \times C)$, these sets are indeed "equivalent" in some way. Indeed, the set that we will soon define as "the" three-fold Cartesian product will also be "equivalent" to these two sets as well, in the sense that there will be an obvious bijective function from the set that we will

SOLUTIONS TO CHAPTER 19 293

define to either of the two sets $(A \times B) \times C$ or $A \times (B \times C)$. In order to make these statements more rigorous, we will need category theory. Thus, let us simply be patient and content without full rigor, and move on to defining what we will call "the" three-fold Cartesian product.

To do this, we must reflect on what the three-fold Cartesian product is really doing for us. Namely, the Cartesian product of A and B is the set of all "pairs" of elements, with one coming from A and one coming from B. Let us similarly define the Cartesian product of three sets A, B, and C, to be the set of all "triples" of elements, with one coming from A, one coming from B, and one coming from C. Thus, just as $A \times B$ is the set of all elements of the form (a, b) with $a \in A$ and $b \in B$, let us **define** the set $A \times B \times C$ (note the lack of parentheses in this notation) to be the set of all elements of the form (a, b, c) with $a \in A$, $b \in B$, and $c \in C$.

Let us see how this works if we take $A = \{1, 2\}$, $B = \{2, 3\}$, and $C = \{1, 3\}$ as given in this exercise. We want to form all possible "triples" with the "first slot" containing an element of A, the "second slot" containing an element of B, and the "third slot" containing an element of C. To make sure that we do not miss any, let us start with the first element[19] of A, the first element of B, and the first element of C. This gives the element $(1, 2, 1)$ of $A \times B \times C$. Let us now stay on the first element of A as well as the first element of B, but then move on to the second element of C. This gives the element $(1, 2, 3)$ of $A \times B \times C$. Let us now stay on the first element of A, go to the second element of B, and go back to the first element of C. This gives the element $(1, 3, 1)$ of $A \times B \times C$. Continuing on in this way shows us that

$$A \times B \times C = \{(1,2,1), (1,2,3), (1,3,1), (1,3,3),$$
$$(2,2,1), (2,2,3), (2,3,1), (2,3,3)\}.$$

It is now clear that there is indeed a bijective function between $A \times B \times C$ and either of the two sets $(A \times B) \times C$ and $A \times (B \times$

---

[19] We note that there is no "first" or "second" element of any set — namely, there is no "order" to these sets. Here, by "first" and "second" we simply are referring to the way in which they are written on the page, from left to right.

C). For example, this function would simply take any element from $(A \times B) \times C$ and send it to the element in $A \times B \times C$ that does not have the internal parentheses. In particular, the element $((1,2),1)$ in $(A \times B) \times C$ would be sent to the element $(1,2,1)$ in $A \times B \times C$.

We immediately see that the set $A \times B \times C$ has all the same **information** as $(A \times B) \times C$ and $A \times (B \times C)$ but without the downside of depending on some arbitrary choice of which sets to take the Cartesian product of first. This is why we call $A \times B \times C$ "the" three-fold Cartesian product, and we must simply keep in the back of our minds that the sets $(A \times B) \times C$ and $A \times (B \times C)$ are in some sense "equivalent" to it, but also that they come with some extra baggage.

As our final comment about this comically lengthy solution, let us note that the definition of the three-fold Cartesian product can indeed be generalized to a four-fold, five-fold, six-fold, and indeed N-fold Cartesian product (where N is **any** positive whole number greater than 1). As an example, if we want the four-fold Cartesian product of the sets A, B, C, and D, we simply consider the set of all "quadruples" of the form (a, b, c, d) where $a \in A$, $b \in B$, $c \in C$, and $d \in D$.

This ends our discussion of many-fold Cartesian products. There will be many contexts in our future in which these ideas (and their analogs in the case of many-fold disjoint unions) are crucial, and we will refer back to this discussion often. Namely, many-fold Cartesian products of sets are absolutely vital for various considerations in geometry, and we will gain more familiarity with these ideas when the time comes. For now, the important thing to take away from this whole discussion is that sometimes in math we make definitions that have some problems with them. In this case, the problem arose from the difficulty of taking many-fold disjoint unions and Cartesian products. However, the beauty of math lies in its flexibility, and we are indeed free to create new ways around these problems (so long as these new methods are still perfectly well-defined and not at all contradictory to the preceding logical framework that has been built). Indeed, mathematics is the language that we use to describe the logical world. And though we are not free to choose what goes on in the logical world (as that world has been set in stone without our involvement), we are indeed free to choose the most convenient ma-

SOLUTIONS TO CHAPTER 19

chinery, or language, with which to explore this world.

2) Let A, B, and C be as in the first exercise above. How many elements are in the set $(A \cup B) \sqcup C$? Explicitly write down an element (any element) from this set. (Recall that this is the disjoint union of the set $A \cup B$ with the set C.)

Solution: We recall again that the disjoint union of a set with N elements and a set with M elements is a set with $N + M$ elements. Namely, if A is a finite set with N elements and if B is a finite set with M elements, then $A \sqcup B$ has $N + M$ elements. Using this, we can quickly see how many elements are in $(A \cup B) \sqcup C$. Namely, C has two elements, and so once we find out how many elements are in $A \cup B$ we will be done — all we will have to do is add two to this number. Now, since $A \cup B$ is the **regular** union of A and B, we do not "double count" the elements that are in both A and B. Namely, since $A = \{1, 2\}$ and $B = \{2, 3\}$, we have that $A \cup B = \{1, 2, 3\}$. Thus, $A \cup B$ has three elements, and therefore $(A \cup B) \sqcup C$ has $3 + 2 = 5$ elements.

Now let us handle the second part of this exercise — namely, let us explicitly write down an element from this set. However, instead of writing down only one element from this set, we will write down all of the elements of this set. Any of these elements would suffice as a solution to this part of the problem.

We recall that in order to write down an element in the set $A \sqcup B$, we must first "label" all of the elements in A with the symbol "A," as well as "label" all of the elements in B with the symbol "B." We do this by placing the elements of the sets in the "first slot" of a pair, and the name of the set in the "second slot" of the same pair. The set of all such pairs is then the disjoint union of A and B.

In our case, we are taking the disjoint union of the set $A \cup B = \{1, 2, 3\}$ with the set $C = \{1, 3\}$. Thus, we must first create the pairs with the elements of $\{1, 2, 3\}$ in the "first slot," and the symbol $A \cup B$ in the second slot. Namely, these three pairs are $(1, A \cup B)$, $(2, A \cup B)$, and $(3, A \cup B)$. We must then do the same for C, and create the pairs $(1, C)$, and $(3, C)$. The set of all of these pairs is then $(A \cup B) \sqcup C$, and so we see that we have

$$(A \cup B) \sqcup C = \{(1, A \cup B), (2, A \cup B), (3, A \cup B), (1, C), (3, C)\}.$$

Explicitly writing down any one of these elements would suffice for this exercise, and we note that we did all of this by simply, calmly, and **strictly** following our definition of a union and a disjoint union.

3) Again let A, B, and C be as in the first exercise above. How many elements are in the set $(A \cap B) \sqcup C$? Explicitly write down an element (any element) from this set.

Solution: We proceed in the same way we did in the previous exercise. Namely, we know that C has two elements and therefore we only need to know how many elements are in $A \cap B$. Since 2 is the only element that is in both A and B, we quickly see that $A \cap B$ has only one element (and it is the number 2). Thus, since the number of elements in $(A \cap B) \sqcup C$ is the sum of the number of elements in $A \cap B$ and the number of elements in C, we see that $(A \cap B) \sqcup C$ has $1 + 2 = 3$ elements.

We now turn our attention to writing these elements down — any one of which would suffice for this exercise. Namely, we know that we must create pairs with elements in $A \cap B$ in the "first slot" and "$A \cap B$" itself in the "second slot," as well as pairs with elements in C in the "first slot" and "C" itself in the second slot. There is, however, only one element in $A \cap B$ — the number 2 — and so the only pair that we have to create from this set is $(2, A \cap B)$. As in the previous exercise, the pairs that we get from C are $(1, C)$ and $(3, C)$. Thus, we see that

$$(A \cap B) \sqcup C = \{(2, A \cap B), (1, C), (3, C)\}.$$

Explicitly writing down any one of these elements would suffice for this exercise, and we note that we did all of this by simply, calmly, and **strictly** following our definition of an intersection and a disjoint union.

4) Yet again let A, B, and C be as in the first exercise above. How many elements are in the set $(A \times B) \sqcup C$? Explicitly

SOLUTIONS TO CHAPTER 19

write down an element (any element) from this set.

Solution: We have now seen how this works a couple of times so let us dive right in. Namely, $A \times B$ has four elements, since A and B each have two elements and the number of elements in the Cartesian product of two sets is the product of the number of elements in the constituent sets. For concreteness, we have that

$$A \times B = \{(1,2), (1,3), (2,2), (2,3)\}.$$

Thus, since C has two elements, we know that $(A \times B) \sqcup C$ has $4 + 2 = 6$ elements.

Let us now turn our attention towards writing all six of these elements down. Namely, we know that we get four pairs from the set $A \times B$, where an element of $A \times B$ goes in the "first slot" and the symbol "$A \times B$" goes in the second slot. This gives the pairs $((1,2), A \times B)$, $((1,3), A \times B)$, $((2,2), A \times B)$, and $((2,3), A \times B)$. We note that nested parentheses are required here because the elements of $A \times B$ are **themselves** pairs, like $(1,3)$. Then, by creating the pairs $(1, C)$ and $(3, C)$ from the set C, we see that we have

$$(A \times B) \sqcup C = \{((1,2), A \times B), ((1,3), A \times B),$$
$$((2,2), A \times B), ((2,3), A \times B), (1, C), (3, C)\}.$$

Explicitly writing down any one of these elements would suffice for this exercise, and we note that we did all of this by simply, calmly, and **strictly** following our definition of a Cartesian product and a disjoint union.

Appendix A: Vacuous Truths

In this appendix we will introduce the notion of a "vacuous statement" — a statement that is true, but completely devoid of meaning. In particular, we will learn what it means for some statement to be "vacuously true." Statements that are vacuously true come up from time to time in mathematics, and their existence (and truth) often are crucial for certain mathematical constructions to be logically sound. For example, we made use of a vacuous truth in Chapter 2, where we proved that the empty set is a subset of every set. This is a great example of a vacuous statement, so let us explore it in detail again.

Let us first remind ourselves briefly about subsets. Recall that if we have some set A, then a subset of A is some set whose every element is also in A. If the enemy hands us some set B and asks us if it is a subset of A, all we need to do is consider every single element in B and ask if it is also an element of A. If the answer is yes for every element in B, then the answer to the enemy's question is yes. Similarly, if there is at least one element in B such that the answer to this question is no — namely, if there is some element in B that is **not** also an element of A — then B is not a subset of A and so the answer to the enemy's question is no.

Suppose, however, that the enemy handed us the empty set and asked if it was a subset of A. What would be our answer? As before, all we would have to do is look at every element of the empty set and ask if it is also an element of A. But there are no elements in the empty set. Thus the statement "every element in the empty set is also in A" is true simply because there are no elements in the empty set to even consider! Moreover, the statement "every element in the empty set is also in A" is an example of a **vacuous truth**. This is because even though this statement is true, it does not actually **mean** anything — namely, we have not actually **learned** anything about any actual elements.

Let us take another example, which will likely make the notion of a vacuous truth even clearer. This example will not be as mathematically precise as the vacuous statement in the previous paragraph, but it will bring out the essential features of vacuous statements and thus make the more mathematically precise ones easier to handle. Consider the following[1] statement: "Whenever there are cows on the moon, we can fly." Now, despite our best wishes and possible experimentation, we know that we cannot fly (at least, we cannot fly without some other assistance, like metal wings and several powerful jet engines). Thus the statement "we can fly" is certainly false.

However, it happens to be true that we **can** fly, provided that there are cows on the moon. Namely, **every time in the history of the universe** that there has been a cow on the moon, we have had the ability to fly. Of course (presumably), there has never been a cow on the moon. Also of course (at least we can be quite sure), we have never been able to fly. Thus it is indeed the case that the presence of cows on the moon has coincided perfectly with the instances of us being able to fly. Accordingly, the statement "whenever there are cows on the moon, we can fly" is indeed true! The reason why this truth is vacuous is that it indeed tells us nothing about when we **actually** can fly.

Now before we go crazy proving statements like "As long as pigs can speak French, Lebron is better than Kobe" (which would be true only in this circumstance), we should always keep in mind that there is a reason why we call these statements "vacuous." Namely, they are devoid of any content (like a vacuum, hence the term[2] "vacuous").

---

[1] The following may sound totally absurd, but it is actually mathematically relevant!

[2] This is a highly unsubstantiated claim. Namely, the author admits that he knows nothing about linguistics, and so cannot verify that we use the term "vacuous" here because it is similar to the word "vacuum." It is just an assumption that the author is making — perhaps a false one — that there is some meaning behind the fact that both "vacuum" and "vacuous" begin with "vac."

Thus, even though vacuous statements are true, they do not tell us anything new. No one will be handed any awards for proving a vacuous statement, simply because they are automatically true and they do not give us any new information.

For example, being able to prove that Lebron is better than Kobe whenever a pig can speak French does **not** mean that we can prove it in any scenario that might actually be relevant to the real world. The truth of the statement just kind of sits "out there" lacking any real substance. Nonetheless, the validity of some vacuous statements is important for certain mathematical results, and certain logical consistency. In particular, we used the fact that the empty set is a subset of every set (even itself) in order to establish the general pattern first described in Chapter 2, and proved in the next appendix. The fact that the empty set is a subset of every set will also be extremely important when we define topologies as well as when we define categories (which will both happen in much later volumes). Thus, despite the fact that the statement "the empty set is a subset of every set" is only vacuously true, its truth is **necessary and important** for much of the logical consistency of mathematics.

We can now have fun beating our friends in arguments by inserting various vacuously true statements in such a way that our argument is untouchable. Once we perfect this craft we can seriously consider a career in politics!

Appendix B: Proving The Pattern

We now set out to prove the general pattern that was conjectured to hold in Chapter 2. Namely, we set out to show that if a set has N elements in it — where N is some finite number — then that set always has exactly 2^N distinct subsets. Although this appendix is primarily intended for mathematical rigor and completeness (since we want to minimize the amount of times we have to "take things on faith" as we did in Chapter 2), it will also provide a good example of how we sometimes need to get creative when we prove certain statements. There are several ways to prove that this pattern holds, and some of these methods require fancier machinery than we currently have available. However, with a little creativity, we can make the result obvious just from the theory of sets, functions, and counting that we developed in the first eight chapters of this volume. Indeed, we do not recommend reading this appendix until the end of Chapter 8 has been reach. The solution to the final exercise in Chapter 5 is a prerequisite for this appendix as well, as we will be using one of the results from that solution in this proof.

Before we begin, however, let us wax philosophical for a bit. In particular, let us (somewhat) briefly discuss what a mathematical proof really is.

The question of what constitutes a mathematical proof is a deep one — one that we will have much more to say about in later volumes and one which has no clear answer. One relatively well agreed upon definition of a proof, and the definition that we will go with here, is a finite sequence of logically deduced and unassailably true statements which demonstrates the validity of a final statement. Now, there are lots of terms in there that are ill-defined and/or contentious, and there is nothing we can do about that here. Let us recall again that this is a huge question — one that we will have no pretention of being able to do justice to here.

One thing we can gather from this definition of a proof, however, is the notion that a proof should in a sense lay a red carpet out for our minds, leading our brains one step at a time through a supremely convincing argument for the validity of some statement, using only previously determined truths and logical manipulations. Thus, when proving a statement it does not matter whether one needs to invoke some high powered theorem about braided monoidal categories[1] or if one can attain the desired result using nothing but basic arithmetic. All that matters in a proof is rigor and clarity — not complexity. In fact, it is often the simple and short proofs that inject the beauty into math, and not the 150 page tours de force that some theorems seem to require for their proof.

The previous two paragraphs are only here to make explicit what we mean by "proving the pattern." Namely, we will provide a sequence of statements that make it clear that a set with N elements must have 2^N distinct subsets. Before beginning, however, we must introduce one small piece of machinery. We have already defined sets and saw that we could just as well define a **set of sets**, whose elements are themselves sets (see Chapter 3 for more on this). Let us now consider **sets of functions**. This is not so bad at all — it is just a set whose individual elements are functions. This is possible since we have a way of distinguishing two functions from each other: we say two functions f and g are the same, or equal, if their domains and codomains are the same, and if $f(a) = g(a)$ for each element a in the domain. This is just the technical way of saying that if the functions act on the same domain in the same exact way, then they are the same functions. We note that there is nothing "written in the stars" that tells us what it should mean for two functions to be equal to each other, and therefore we need to **define** our own notion of function equality. The notion that we just introduced — that two functions are equal if they act

---

[1] We are not yet meant to understand what these words mean — they are purposefully chosen so as to sound fancy.

between the same domain and codomain in the exact same way—is the most obvious definition of equality of functions. Let us see an example of a set of functions. Consider the sets $A = \{1, 2\}$ and $B = \{a, b\}$. There are only 4 possible functions that we can define from A to B, since we can have

$$1 \to a \quad \text{and} \quad 2 \to a,$$

or

$$1 \to a \quad \text{and} \quad 2 \to b,$$

or

$$1 \to b \quad \text{and} \quad 2 \to a,$$

or

$$1 \to b \quad \text{and} \quad 2 \to b.$$

Any function from A to B must be equal to one of these functions, so if we label these four functions as f_1, f_2, f_3, and f_4 then we can form the set $\{f_1, f_2, f_3, f_4\}$ of functions from A to B.

Now we are ready to prove the pattern that a set with N elements has 2^N distinct subsets. We must first understand how all of this "sets of functions" business relates to the number of distinct subsets of a set. Indeed, these two concepts are related in a very natural way. For let us let A be a set with N elements and let B be a subset of A (any subset at all). We can then fully characterize B by considering a clever function from A to the set $\{0, 1\}$ as follows. Let us define the function from A to $\{0, 1\}$ by having it send an element in A to 0 if that element is **not** in B, and to 1 if that element **is** in B. Let us denote this function by f_B to remind ourselves that this function depends heavily on the subset B. Thus, the elements in B are simply the elements in A that get sent to 1 by the function f_B, and all other elements in A (i.e., those that are not in B) get sent to 0 by the function f_B.

Since we were able to define the function f_B regardless of what the details of the subset B are, we see that we can define this function for **any** subset of A that we would like. As an example, if we chose the empty set as our subset of A so that $B = \emptyset$, then the function f_B would simply be the function that maps **every element** in A to

0. This is because f_B maps only those elements in A that are not in B to 0, and since nothing in A is in the empty set, we see that every element in A gets sent to 0 by f_B when B is the empty set. As a second example, let us consider the case when our subset B is the set A itself, and let us ask what the corresponding function f_B is. A little thought shows that when $B = A$, the corresponding function f_B is just the function that sends every element in A to 1, since now every element in A is in the subset B and since f_B sends the elements in A that are in B to 1.

It is clear that if we are handed a subset of A, then we can construct the function that sends the elements in A that are not in this subset to 0 and the elements in A that are in the subset to 1. Moreover, we see that if instead we are handed a function $f : A \to \{0,1\}$, then we can define a subset of A by simply taking the elements that get sent to 1 by this function. Thus, there is a bijective correspondence between subsets of A and the functions $f : A \to \{0,1\}$ — namely, for every subset of A there is one and only one function from A to $\{0,1\}$ (which maps elements in the subset to 1 and elements not in the subset to 0), and for every function from A to $\{0,1\}$ there is one and only one subset of A.

As an example of all of this, let us consider the set $A = \{a, b, c\}$. We have already seen that this set has eight subsets, so let us see how these subsets correspond to the eight functions from A to $\{0,1\}$ in the manner we just described. The empty subset corresponds to the function (which we will choose to denote by f_1) that sends all of the elements in A to 0. The subset $\{a\}$ corresponds to the function (which we will choose to denote by f_2) that sends a to 1 and both b and c to 0 (since a is in this subset but b and c are not). The subset $\{b\}$ corresponds to the function (which we will choose to denote by f_3) that sends b to 1 and both a and c to 0. The subset $\{c\}$ corresponds to the function (which we will choose to denote by f_4) that sends c to 1 and both a and b to 0. The subset $\{a, b\}$ corresponds to the function (which we will choose to denote by f_5) that sends both a and b to 1 and c to 0. The subset $\{a, c\}$ corresponds to the function (which we will choose to denote by f_6) that sends both a and c to 1 and b to 0. The subset $\{b, c\}$ corresponds to the function (which we will choose to denote by f_7) that sends both b and c to 1 and a to 0.

Finally, the subset $\{a, b, c\}$ (i.e., A itself) corresponds to the function (which we will choose to denote by f_8) that sends a, b, and c all to 1. Indeed, there are no other possible functions from A to $\{0, 1\}$ (as one can explicitly check), and thus there are 8 functions from A to $\{0, 1\}$, just as there are 8 subsets of A.

We have therefore seen that if we can count the number of distinct functions from A to $\{0, 1\}$, then we will have counted the number of distinct subsets of A. However, this is precisely what we proved in the solution to the final exercise of Chapter 5. Namely, in the solution to Exercise 4 (b) in Chapter 5, we showed that if A has N elements, then there are 2^N functions from A to $\{0, 1\}$. We therefore will not prove this result again here, but rather we will simply quote this result and use it for our present purposes. Namely, we have already shown that if a set A has N elements, then the number of distinct subsets of A is equal to the number of distinct functions from A to $\{0, 1\}$. By quoting the result that the number of distinct functions from A to $\{0, 1\}$ is 2^N when A has N elements, we have successfully shown that the number of distinct subsets of A is 2^N when A has N elements. This completes our proof.

Nice Job!

This was a rather tedious proof,[2] but it shows how we are at liberty in interpreting our mathematical structures, so long as these interpretations are completely rigorous. Namely, the key to the proof was realizing that we can associate subsets to certain functions in a bijective way. This association then made the counting a lot more obvious, and more importantly **it allowed us to use a result that we had previously established**. This is a very common and fruitful method of solving math problems and/or proving mathematical theorems — namely, if we can shake the given problem into the form of a problem that we have already solved, then the solution becomes trivial. By relating the number of subsets of a given set to the number of functions from that set to the set $\{0, 1\}$, we not only cracked the

---

[2]Especially when we include the work that we had to do in the solution to Exercise 4 (b) in Chapter 5.

problem wide open by being able to call upon a previous result, but we were also able to make a connection between two seemingly disparate areas of math — subsets and functions. This is one example of one of the most beautiful aspects of math — namely, the ability for connections to be made between structures and/or patterns that were previously thought to be completely disconnected. We will see many more examples of this as we continue on through these volumes, and the more complex and (seemingly) disconnected our mathematical structures become, the more remarkable the connections amongst them become.

Let us also note that the proof given in the previous section helps to clarify **why** the statement under consideration is true. This is another sign of mathematical beauty. Namely, if a proof simply involves quoting some high-powered mathematical statement from some unrelated sub-field of math, then it is neither helpful, enlightening, nor beautiful. Such a proof would indeed verify the validity of the statement that we were concerned with, but it would leave us in the dark about the very essence of what the truth is trying to tell us. However, if a proof can establish the validity of the statement we are concerned with **as well as** help to make clear **why** this statement is true, and why it could not be any other way, then we can see the true beauty in math directly — namely, we can see how pure logic drives us into the inevitable and eternal truths that we are so fortunate to experience.

Index

axiom, 112

cardinality, *see* set
Cartesian product, 185
codomain, *see* function

Descartes, 189
disjoint union, 190
domain, *see* function

element, 15
empty set, 23
Euclid, 117

fraction, 96
function, 48
    codomain of, 53
    bijective, 63
    domain of, 53
    injective, 58
    sets of, 302
    surjective, 61

Georg Cantor, 132

Hilbert Hotel, 70

integer, 217
intersection (of sets), 162

natural numbers, 67

power set, 35, 150
prime numbers, 117
    proof of infinitely many, 121
proof by contradiction, 110

rational number, 217
real number, 126
Russell's paradox, 43

set, 15
    cardinality of, 80, 86
    empty, 23
    sets of, 32
    sub-, 20
subset, 20
    proper, 216

union (of sets), 162

Index Of Notation

Symbol(s)	Description	Chapter Introduced
$A = \{...\}$	A is the set of elements $\{...\}$ contained in the brackets.	1
$P(\cdot)$	The power set.	3
$f : A \to B$	f is a function from A to B.	5
$f(a) = b$	The element a is sent to the element b by the function f.	5
$\mathbb{N}$	The set $\{1, 2, 3, ...\}$ of positive whole numbers.	7
$\mathbb{R}$	The set of all real numbers.	14
$\Rightarrow$	"$P \Rightarrow Q$" reads "P implies Q."	15
$\neg$	"$\neg P$" reads "not P."	15
$\cup$	$A \cup B$ is the union of A and B.	16
$\cap$	$A \cap B$ is the intersection of A and B.	16
$\in$	$a \in A$ means a is an element of A.	16
$\{...|...\}$	The set of elements ... "such that" the condition ... is satisfied.	16
$\subseteq$	$B \subseteq A$ says that B is a subset of A.	17
$\times$	$A \times B$ is the Cartesian product of A and B.	18
$\sqcup$	$A \sqcup B$ is the disjoint union of A and B.	19
$\subset$	$B \subset A$ says that B is a proper subset of A.	17
$\mathbb{Z}$	$\{..., -3, -2, -1, 0, 1, 2, 3, ...\}$	21
$\mathbb{Q}$	The set of all rational numbers (i.e., fractions).	21

Made in United States
Orlando, FL
06 February 2023